ZHA – SUSTAINABLE AI FRAMEWORK FOR AI TRANSFORMATIONS

DESIGNING AND TRANSFORMING THE HUMAN-CENTRIC AI ORGANIZATION TO CIVILIZATION

KARKI ASHOKKUMAR

Made with ♥ on the Notion Press Platform
www.notionpress.com

Zha Sustainability Practitioners Research Series – Volume 3

Contents

Contents

Preface

Humanity stands at a defining threshold in its evolutionary journey.

Artificial Intelligence is no longer a technological experiment confined to laboratories or innovation centers. It has become the invisible infrastructure shaping economies, governance, education, agriculture, healthcare, relationships, and increasingly, human identity itself. Decisions once made solely by individuals are now influenced, augmented, and sometimes determined by intelligent systems.

As organizations across the world embrace AI transformation, conversations around Responsible AI have emerged to address ethical risks, fairness, transparency, privacy, and accountability. These principles are essential safeguards.

Yet through years of observing technology adoption, enterprise transformation, sustainability initiatives, and societal behavior, a deeper realization emerged:

Responsible AI protects. Sustainable AI guides.
Protection manages risk. Guidance shapes destiny.

This insight led to the evolution of the ZHA Sustainable AI Framework, a new paradigm that moves AI transformation beyond governance toward civilizational design.

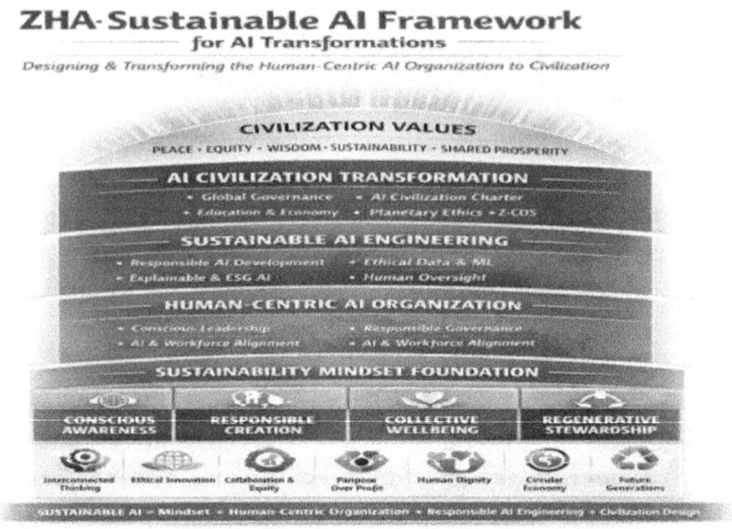

The Evolution of the ZHA Research Journey

This book represents Volume 3 of a continuing body of research and development emerging from the ZHA Foundation ecosystem.

Volume 1 - ZHA Methodology for Sustainability Mindset
established the philosophical and behavioural foundation. It explored how human consciousness, values, and decision-making patterns determine the sustainability of societies, institutions, and innovations. It introduced the core principle that sustainable systems begin with sustainable thinking.

Volume 2 - ZHA Sustainability Practitioners Certifications
translated philosophy into practice. It operationalized the Sustainability Mindset through structured certifications, gamified learning ecosystems, institutional governance models, and real-world implementation across

schools, communities, and organizations — transforming individuals into sustainability practitioners.

Volume 3 - ZHA Sustainable AI Framework for AI Transformations extends this evolution into the age of Artificial Intelligence.

If Volume 1 shaped mindset, and Volume 2 enabled practice, Volume 3 designs the future of intelligent civilization.

This volume scales the research from human transformation to organizational transformation and ultimately toward civilization transformation.

Beyond Responsible AI

Responsible AI often enters the conversation after technology is created — functioning as a corrective governance layer.

However, many modern disruptions do not originate from unethical algorithms alone. They emerge from deeper systemic imbalances:

- Over-optimization for profit without societal balance
- Innovation disconnected from planetary wellbeing
- Technology developed without long-term thinking
- Product engineering driven by speed rather than purpose
- Organizations transforming digitally without transforming culturally

Enterprises frequently begin AI transformation by selecting tools, architectures, or methodologies:

Agile versus Waterfall.
DevOps versus traditional delivery.
Automation pipelines, data platforms, or AI models.

Yet the most fundamental question is rarely asked:

What mindset should guide the creation of intelligence itself?

Before selecting an SDLC model...
Before designing a product...
Before deploying AI systems...

Human intention must align with sustainability and purpose.

Without this alignment, even advanced systems can accelerate inequality, environmental strain, institutional fragility, and societal anxiety.

The Birth of the ZHA Sustainability Mindset

The ZHA Sustainability Mindset did not originate as an academic theory. It evolved through lived experimentation between 2023 and 2026, engaging students, professionals, farmers, educators, enterprises, and communities.

One insight became undeniable:

Technology problems are often human mindset problems expressed through machines.

AI amplifies human intention at scale.

If human thinking is driven by scarcity, competition, and short-term gain, intelligent systems magnify those patterns globally.

But when thinking evolves toward stewardship, collaboration, sufficiency, and long-term vision, AI becomes a force for collective wellbeing.

The ZHA Sustainability Mindset introduced foundational principles guiding individuals and institutions to balance:

- Progress with preservation
- Intelligence with wisdom
- Innovation with responsibility
- Growth with harmony

Sustainable AI - The New Benchmark for AI Transformation

This book proposes a decisive shift:

Sustainable AI must become the starting point of AI transformation - not its correction mechanism.

The ZHA Sustainable AI Framework integrates:

- Human dignity
- Environmental stewardship
- Social equity
- Ethical intelligence
- Long-term civilization thinking
- Continuous adaptive learning

Under this paradigm:

SDLC methodologies are selected based on sustainability outcomes.

Product development begins with human purpose.

AI governance becomes proactive rather than reactive.

Organizations evolve from productivity engines into stewardship institutions.

Technology aligns with life itself.

AI transformation is therefore reframed, not as digital modernization, but as civilizational evolution.

From Human-Centric Organizations to AI Civilization

The journey documented in this volume connects multiple layers of transformation:

- Sustainability Mindset Philosophy
- Human-Centric AI Organization Design
- Sustainable AI Transformation Framework
- AI Product & SDLC Alignment Models
- Governance and Global Policy Thinking

ZHA Global Think Tank & World Summit Vision

ZHA 2035 Civilization Roadmap

The ZHA Civilization Operating System (Z-COS)

Together, they form the ZHA Sustainable AI Civilization Framework — a blueprint for designing institutions capable of thriving in the age of intelligent systems.

This work is not written solely for technologists.

It is written for:

- Leaders redesigning organizations
- Policymakers shaping governance
- Educators preparing future generations
- Innovators building intelligent products
- Citizens seeking meaning in an AI-driven world

Because the future of Artificial Intelligence is inseparable from the future of humanity.

A Personal Conviction

Humanity does not merely need smarter machines.

Humanity needs wiser intelligence.

The ZHA Sustainable AI Framework represents a commitment to ensure technology remains a servant of life, never its master.

It invites us to design systems that:

- nurture communities,
- protect the planet,
- elevate human consciousness, and
- enable future generations to inherit not only advanced technologies but a thriving civilization.

This book is an invitation:

An invitation to rethink innovation.
An invitation to redesign transformation.
An invitation to co-create a conscious AI civilization.

By, Karki Ashokkumar
Founder & Mentor, ZHA Foundation Charitable Trust
Transformation Architect & Author, ZHA Sustainable AI & AI Civilization Framework

PART I - Awakening: Humanity at a Turning Point

CHAPTER I

Chapter 1: The Age of Technological Disruption

This chapter explores the global challenges caused by uncontrolled technological growth and introduces the foundational thinking behind Sustainable AI with Human-Centric Purpose.

1.1 Humanity at an Inflection Point

Every civilization reaches moments when progress accelerates faster than wisdom.

- The Industrial Revolution transformed human labor.
- The Digital Revolution transformed information.
- Today, Artificial Intelligence is transforming human decision-making itself.

We are entering an era where machines do not merely assist humans; they increasingly predict, recommend, automate, and influence human choices. Algorithms curate knowledge. Autonomous systems manage infrastructure. Intelligent platforms shape economies. Generative AI begins to participate in creativity, reasoning, and governance.

This transformation represents one of the most powerful technological shifts in human history.

Yet alongside remarkable innovation, a silent question emerges:

Is humanity evolving at the same pace as its technology?

The Age of Technological Disruption is not defined only by innovation. It is defined by imbalance — the widening gap between technological capability and human readiness.

1.2 Technology Without Direction

Technological growth has historically been celebrated as progress. Faster computing, automation, data intelligence, and global connectivity have improved efficiency and expanded opportunity.

However, modern disruption differs from earlier revolutions in one critical way:

Technology now scales human intention globally and instantly.

When intention is constructive, technology amplifies prosperity.
When intention lacks balance, technology magnifies risk.

Across the world, several patterns have begun to emerge:

Acceleration Without Reflection

Organizations rush to adopt AI systems without understanding long-term consequences. Innovation cycles prioritize speed over sustainability.

Optimization Without Meaning

Algorithms optimize engagement, productivity, and profit but rarely consider psychological well-being, environmental impact, or social harmony.

Intelligence Without Wisdom

AI systems grow increasingly capable, while societies struggle with polarization, misinformation, anxiety, and ecological strain.

Technology itself is not the problem.
The absence of guiding philosophy is.

1.3 The Illusion of Infinite Growth

Modern technological ecosystems operate on an assumption inherited from industrial economics:

Growth must always accelerate.

- Startups scale rapidly.
- Enterprises automate aggressively.
- Digital platforms pursue endless engagement.

While growth drives innovation, unchecked growth introduces systemic vulnerabilities:

- Resource consumption expands faster than planetary regeneration.
- Automation reshapes employment faster than societies can adapt.
- Digital dependency increases faster than psychological resilience develops.

The result is a paradox:

Humanity has never been more technologically advanced, yet uncertainty, burnout, environmental stress, and institutional fragility continue to rise.

Technological disruption exposes a deeper civilizational challenge:

Progress measured only by capability eventually disconnects from well-being.

1.4 AI as the Ultimate Disruptor

Artificial Intelligence represents a qualitative shift unlike previous technologies.

- Earlier tools extended human strength.
- Digital systems extended human memory.
- AI extends human cognition.

For the first time in history, humanity is creating systems capable of learning, reasoning, generating knowledge, and influencing collective behavior.

This introduces unprecedented opportunities:

- Medical breakthroughs
- Climate modeling
- Educational accessibility
- Intelligent infrastructure
- Acceleration of scientific discovery

But it also introduces unprecedented risks:

- Algorithmic bias embedded at scale
- Autonomous decision-making without accountability
- Concentration of technological power
- Erosion of human agency
- Societal dependence on opaque systems

AI is neither inherently beneficial nor dangerous. It becomes what human civilization programs it to become.

1.5 Disruption Beyond Technology

The disruption we face is not purely technological — it is systemic.

Economic Disruption
Automation transforms industries faster than workforce reskilling occurs.

Social Disruption
Digital platforms reshape relationships, communication, and identity.

Institutional Disruption
Governance structures struggle to regulate technologies evolving exponentially.

Psychological Disruption

Humans confront existential questions about purpose, creativity, and relevance in an automated world.

Technology exposes vulnerabilities already present in human systems.

AI does not create civilization problems; it reveals them.

1.6 The Limitation of "Responsible AI"

In response to emerging risks, global discussions have introduced the concept of Responsible AI.

Responsible AI focuses on:

- Fairness
- Transparency
- Accountability
- Privacy
- Ethical safeguards

These principles are essential.

However, Responsible AI often functions as a reactive framework applied after technology has already been designed.

It asks:

- Is the system fair?
- Is it compliant?
- Is harm minimized?

But it rarely asks:

- Should this system exist?
- Does it serve humanity's long-term well-being?
- Does it strengthen civilization or weaken it?

Responsible AI governs behavior.
It does not define purpose.

And without purpose, responsibility becomes insufficient.

1.7 The Missing Element: Mindset
Through observing technological adoption across organizations, education systems, and communities, a consistent pattern emerges:

Technological failures often originate from human mindset limitations.

When innovation is driven by:

- Fear of competition
- Short-term profit
- Prestige
- Unchecked ambition

Technology accelerates imbalance.

Conversely, when innovation emerges from:

- Sufficiency thinking
- Collaboration
- Stewardship
- Long-term vision

Technology becomes regenerative.

The true foundation of sustainable innovation is therefore not hardware, software, or algorithms.

It is human mindset.

This insight became the origin of the ZHA Sustainability Mindset.

1.8 Introducing Sustainable AI with Human-Centric Purpose

The Age of Technological Disruption demands a new benchmark.

Not faster AI.
Not merely responsible AI.

But Sustainable AI with Human-Centric Purpose.

Sustainable AI shifts the starting point of innovation.

Instead of asking:
"How powerful can technology become?"

We ask:
"How beneficial can technology be for humanity and the planet?"

This framework introduces three essential alignments:

Human Alignment
Technology must enhance dignity, creativity, and human flourishing.

Planetary Alignment
Innovation must operate within ecological limits and support environmental regeneration.

Civilizational Alignment
AI must strengthen institutions, communities, and future generations.

Under this model, sustainability is not an afterthought — it becomes the design principle guiding transformation.

1.9 Rethinking Transformation Before SDLC and Product Development

Organizations traditionally begin transformation by selecting methodologies:

- Agile
- DevOps
- Waterfall
- Hybrid SDLC models
- Data architectures
- Automation pipelines

Yet methodology selection occurs too late in the transformation journey.

Before choosing an SDLC...
Before defining product roadmaps...
Before deploying AI systems...

Organizations must establish:

Why are we building this intelligence?

Sustainable AI reframes transformation:

- SDLC becomes a tool aligned with purpose.
- Product development becomes an act of stewardship.
- Innovation becomes a responsibility toward society.

Technology follows mindset — not the other way around.

1.10 From Disruption to Conscious Civilization

Every technological era forces humanity to redefine itself.

The current disruption presents humanity with a choice:

We can allow technology to shape civilization unintentionally or we can consciously design a future where intelligence serves life.

The ZHA Civilization Framework emerges from this conviction:

Humanity must evolve from a technology-driven civilization to a purpose-driven civilization.

The Age of Technological Disruption is therefore not a crisis.

It is an awakening.

An opportunity to rethink progress.
An opportunity to realign innovation.
An opportunity to build a conscious civilization guided by Sustainable AI with Human-Centric Purpose.

Chapter Reflection

Technological disruption is inevitable.
Civilizational decline is not.

The future of AI will not be decided solely by engineers or algorithms.

It will be determined by the mindset humanity chooses to adopt.

The journey toward Sustainable AI begins not with machines — but with human awareness.

CHAPTER II

Chapter 2: Challenges in AI, Over-Greed, and Civilization Risk

This chapter explores the global challenges caused by uncontrolled technological growth and introduces the foundational thinking behind Sustainable AI with Human-Centric Purpose.

2.1 The Hidden Driver Behind Technological Disruption

Artificial Intelligence is often described as humanity's greatest technological achievement. Governments invest billions into AI infrastructure. Corporations compete to dominate AI markets. Nations measure progress through computational power, data accumulation, and innovation speed.

Yet beneath this global race lies a less visible force.

Technology itself is neutral.

The true driver behind disruption is not Artificial Intelligence.

It is human motivation amplified through Artificial Intelligence.

Throughout history, civilizations have not collapsed because of technological weakness. They declined when human ambition lost balance, when expansion exceeded wisdom, when consumption exceeded responsibility, and when power overshadowed purpose.

Today, AI has become the most powerful amplifier ever created for human intention.

If intention is wise, AI becomes a tool for civilization advancement.

If intention is driven by over-greed, AI becomes a systemic risk.

2.2 Understanding Over-Greed

Greed, in its simplest form, is the desire for more.

Human progress has often depended on ambition, the pursuit of improvement, innovation, and growth. Ambition fuels discovery and advancement.

However, over-greed emerges when growth disconnects from necessity and wellbeing.

Over-greed manifests when:

Success becomes accumulation without purpose, competition replaces collaboration, efficiency replaces empathy, speed replaces reflection.

In the AI era, over-greed transforms into a structural phenomenon rather than an individual behavior.

Organizations pursue:

Unlimited data collection, maximum automation, perpetual engagement,accelerated profit cycles.

The objective subtly shifts from serving humanity to dominating markets.

Artificial Intelligence does not create over-greed, it scales it globally.

2.3 The Algorithmic Economy

Modern digital economies operate on algorithms designed to optimize outcomes.

Optimization itself is not harmful. But optimization without ethical boundaries creates unintended consequences.

Algorithms are often trained to maximize:

User engagement, revenue generation, operational efficiency, predictive accuracy.

When these objectives lack human-centric constraints, systems begin optimizing behaviors that may harm individuals and societies.

Examples already visible across the world include:

Social platforms amplifying emotional polarization because conflict drives engagement, automated systems prioritizing productivity over worker wellbeing, recommendation engines reinforcing addiction patterns,

AI-driven markets accelerating inequality through data advantage.

The danger is subtle:

- AI does not decide what is valuable.
- It learns what humans reward.
- If society rewards attention over truth, profit over sustainability, or growth over balance, AI faithfully amplifies those priorities.

Thus, civilization risk emerges not from intelligence itself but from misaligned incentives embedded within intelligent systems.

2.4 The Acceleration Trap

AI introduces exponential acceleration into human systems.

Decision cycles shorten.

Innovation timelines compress.

Market disruptions intensify.

While acceleration increases efficiency, it reduces humanity's ability to pause and evaluate consequences.

This creates what can be called the Acceleration Trap:

New technology emerges.

Competitive pressure forces rapid adoption.

Ethical reflection lags behind implementation.

Societal impacts appear only after disruption occurs.

Organizations often fear slowing down because competitors may advance faster. As a result, collective caution disappears even when individuals recognize risks.

Civilizations historically progressed through gradual adaptation. AI challenges this model by transforming society faster than cultural and institutional learning can occur.

Without intentional guidance, acceleration becomes instability.

2.5 AI and the Concentration of Power

Another civilization risk introduced by AI is the concentration of influence.

Unlike previous industrial tools, AI benefits disproportionately from scale:

More data produces better models, better models attract more users, more users generate more data.

This feedback loop concentrates technological capability in the hands of a few organizations or nations.

When intelligence infrastructure becomes centralized, several risks arise:

Reduced democratic participation in technological decisions, economic imbalance, dependency on opaque systems, erosion of institutional autonomy.

Civilization stability depends on distributed empowerment.

Sustainable AI therefore requires governance models that ensure technology strengthens society collectively rather than consolidating power narrowly.

2.6 The Psychological Cost of Intelligent Systems

Technological disruption is often measured economically or politically, but one of its most profound impacts is psychological.

As AI begins to perform tasks once considered uniquely human—writing, designing, reasoning, diagnosing—individuals confront existential questions:

What is human value in an automated world?

What does meaningful work mean?

How does identity evolve when intelligence is shared with machines?

Without philosophical grounding, technological advancement can produce anxiety rather than empowerment.

Human beings do not seek efficiency alone; they seek purpose.

Over-greed drives systems toward productivity metrics, while Sustainable AI recognizes that civilization stability depends on preserving human dignity and meaning.

2.7 Civilization Risk: A Systemic Perspective

Civilization risk is not a single catastrophic event. It emerges gradually through interconnected pressures

Environmental degradation, economic inequality, institutional mistrust, technological dependency, loss of shared values.

AI intersects with all these dimensions simultaneously.

When guided by short-term incentives, AI accelerates fragmentation.

When guided by sustainability mindset principles, AI becomes civilization stabilizing infrastructure.

The question is not whether AI will transform civilization.

The question is whether civilization will transform itself in time to guide AI wisely.

2.8 Why Responsible AI Is Not Enough

Responsible AI frameworks attempt to reduce harm by introducing ethical guidelines and regulatory oversight.

They are necessary safeguards.

However, Responsible AI addresses consequences rather than origins.

It asks:

Is the algorithm fair?

Is bias reduced?

Is transparency ensured?

But civilization risk originates earlier, during intention formation.

If systems are designed within cultures driven by over-greed, responsibility mechanisms operate as corrective patches rather than foundational solutions.

Sustainable AI shifts the paradigm by redefining success itself.

Instead of asking how to control harmful outcomes, Sustainable AI asks how to design systems whose very purpose promotes collective wellbeing.

2.9 Sustainable AI as an Antidote to Over-Greed

Sustainable AI introduces a counterbalance to the forces driving disruption.

It integrates:

Sufficiency over excess, stewardship over exploitation, collaboration over domination, long-term value over short-term gain.

This approach does not reject innovation or economic growth. Rather, it aligns growth with planetary and human sustainability.

Within Sustainable AI:

Technology serves societal resilience, organizations become guardians of future generations, innovation becomes regenerative rather than extractive.

The ZHA Sustainability Mindset provides the behavioral foundation enabling this shift.

Transformation begins not with regulation but with consciousness.

2.10 Reframing Progress

Civilizations often equate advancement with technological sophistication.

But true progress must answer three questions:

Does technology improve human wellbeing?

Does it protect the planet?

Does it strengthen future generations?

If the answer to any of these questions is uncertain, progress requires reconsideration.

The Age of AI invites humanity to redefine success — not as dominance over nature or markets, but as harmony between intelligence, society, and environment.

2.11 The Turning Point

Human civilization now stands at a turning point.

AI can either:

Accelerate division, over-consumption, and instability, or enable cooperation, sustainability, and conscious evolution.

The deciding factor will not be algorithms.

It will be mindset.

Recognizing over-greed as a systemic risk allows humanity to move toward a higher stage of innovation, one guided by Sustainable AI with Human-Centric Purpose.

Chapter Reflection

Over-greed transforms intelligence into risk.

Purpose transforms intelligence into civilization progress.

Artificial Intelligence does not determine humanity's future.

Human values do.

CHAPTER III

Chapter 3: Why Humanity Needs a Sustainability Mindset

This chapter explores the global challenges caused by uncontrolled technological growth and introduces the foundational thinking behind Sustainable AI with Human-Centric Purpose.

3.1 The Real Crisis of Humanity

When societies discuss global challenges, they often focus on visible symptoms:

- Climate change
- Economic inequality
- Technological disruption
- Social instability
- Mental health crises
- Institutional distrust

Solutions are typically proposed through policy reforms, technological innovation, or economic restructuring.

Yet despite unprecedented scientific advancement, many of these problems continue to intensify.

This raises a fundamental question:

What if humanity's greatest crisis is not technological, economic, or environmental — but cognitive

Human civilization has mastered external development but neglected internal evolution.

We have advanced tools faster than we have advanced thinking.

The result is a civilization rich in capability yet uncertain in direction.

3.2 Progress Without Awareness

Modern civilization celebrates progress through measurable outputs:

- GDP growth
- Productivity
- Technological breakthroughs
- Consumption expansion
- Market valuation

These metrics reward expansion, acceleration, and competition.

However, progress measured solely through expansion creates unintended consequences:

Natural resources are consumed faster than replenished, Innovation outpaces ethical reflection, Societies become efficient but disconnected Individuals experience success without fulfillment, Humanity has learned how to grow, but not always how much growth is enough.

The absence of this awareness lies at the root of civilizational imbalance.

3.3 Technology Mirrors Human Mindset

Artificial Intelligence reveals a profound truth:

Technology does not create behavior — it reflects human behavior at scale.

If societies value cooperation, AI strengthens collaboration.
If societies reward outrage, AI amplifies division.
If markets prioritize profit above well-being, AI accelerates exploitation.

AI systems learn from human data — and human data reflects human mindset.

Therefore, technological transformation cannot succeed without mindset transformation.

The sustainability challenge is fundamentally psychological before it becomes environmental or technological.

3.4 The Origin of Unsustainability

Unsustainable systems rarely emerge from malicious intent.

They arise gradually through ordinary decisions influenced by unconscious assumptions:

- More consumption equals success
- Speed equals progress
- Competition ensures survival
- Efficiency outweighs empathy

Over time, these assumptions become embedded in institutions, education systems, economic models, and organizational cultures.

Civilization begins optimizing for growth without understanding its limits.

This condition creates what may be called collective unconscious overreach — humanity expanding beyond ecological, social, and psychological balance.

Sustainability therefore cannot be achieved only through regulations or technologies.

It requires a transformation in how humans perceive success, value, and responsibility.

3.5 What Is a Sustainability Mindset?

A Sustainability Mindset is not merely environmental awareness.

It is a way of thinking that aligns human actions with long-term harmony between:

- People
- Planet
- Prosperity
- Partnership

It represents a shift from extraction to stewardship.

A person or organization operating with a sustainability mindset asks:

- What impact will this decision have beyond immediate benefit?
- Does this innovation support future generations?
- Are resources being used responsibly?
- Does growth strengthen collective well-being?

Sustainability becomes an everyday decision-making framework rather than a specialized domain.

3.6 The Birth of the ZHA Sustainability Mindset

The ZHA Sustainability Mindset emerged from observing diverse communities — students, educators, professionals, farmers, entrepreneurs, and institutions — navigating modern challenges.

A recurring insight appeared:

People were not resistant to sustainability; they lacked a structured framework to practice it consistently.

Sustainability needed to move from abstract ideals to practical behavioral principles.

Thus, twelve guiding principles were developed to help individuals and organizations internalize sustainable thinking.

These principles address human psychology, social cooperation, innovation ethics, and environmental stewardship simultaneously.

They do not restrict ambition; they refine it.

3.7 The Twelve Sustainability Mindset Principles

The ZHA Sustainability Mindset is built upon twelve interconnected principles:

1. Feeling Sufficient

Encourages mindful resource use and freedom from excessive consumption.

2. Encouraging & Rewarding

Recognizes contributions while individuals are alive, promoting intrinsic motivation.

3. Long-Term Thinking

Evaluates decisions through generational impact rather than short-term gain.

4. Risk-Based Decision Making

Integrates environmental and social risk awareness into innovation strategies.

5. Gender Equality

Promotes inclusive participation and balanced societal development.

6. Sharing & Caring Collaboration

Strengthens collective intelligence through cooperation.

7. Environmental Stewardship

Protects ecosystems as essential partners in civilization survival.

8. Fail Fast for Vision

Encourages learning through experimentation while maintaining ethical direction.

9. Community Engagement

Ensures local voices shape development decisions.

10. Social Equality

Promotes fairness in opportunity and resource distribution.

11. Continuous Learning

Builds adaptive societies capable of evolving responsibly.

12. Resource Conservation

Prioritizes efficiency, reuse, and circular sustainability practices.

Together, these principles form a behavioral operating system for sustainable civilization.

3.8 Sustainability Mindset and Artificial Intelligence

AI transformation fails when treated purely as a technical upgrade.

Organizations frequently implement AI to:

- Reduce costs

- Increase productivity
- Outperform competitors

While valid objectives, they remain incomplete.

Without sustainability mindset alignment, AI accelerates existing systemic weaknesses.

Sustainable AI begins by redefining intent:

- AI must augment human capability, not replace human meaning.
- Automation must support dignity, not dependency.
- Intelligence must protect ecosystems, not exploit them.
- Thus, mindset becomes the foundation upon which ethical AI, responsible governance, and sustainable innovation naturally emerge.

3.9 From Individual Mindset to Organizational Culture

Mindset transformation does not remain personal; it reshapes institutions.

Organizations adopting sustainability mindset principles begin to evolve:

- Leadership shifts from control to stewardship.
- Innovation prioritizes long-term resilience.
- Employees experience purpose alongside productivity.
- Decision-making integrates social and environmental impact.
- Culture becomes aligned with sustainability rather than compliance-driven initiatives.
- Transformation becomes organic rather than forced.

3.10 Education and the Future Generation

Perhaps the most critical application of sustainability mindset lies in education.

Future generations will inherit intelligent technologies far more powerful than those available today.

Teaching technical skills alone will be insufficient.

Students must learn:

- Ethical reasoning
- Collaborative thinking
- Ecological awareness
- Emotional intelligence
- Systems thinking

Education must prepare not only skilled professionals, but responsible civilization builders.

The ZHA Sustainability Practitioners model demonstrates how mindset education can be gamified, measurable, and scalable.

3.11 Sustainability Mindset as Civilization Evolution

Civilizations evolve through stages:

- Survival-driven societies
- Growth-driven societies
- Knowledge-driven societies
- Conscious societies

Humanity now transitions toward the fourth stage.

A conscious civilization understands that intelligence without sustainability threatens its own existence.

The Sustainability Mindset represents humanity's transition from unconscious progress to intentional evolution.

3.12 The Foundation for Sustainable AI with Human-Centric Purpose

Sustainable AI cannot exist without sustainable humans.

Technology inherits the philosophy of its creators.

By embedding sustainability mindset principles into leadership, education, governance, and innovation processes, AI transformation becomes aligned with human flourishing.

The ZHA framework therefore begins not with algorithms or systems architecture, but with awareness.

Mindset precedes methodology.
Purpose precedes productivity.
Human consciousness precedes artificial intelligence.

Chapter Reflection

Humanity does not need less technology.

Humanity needs wiser thinking guiding technology.

The Sustainability Mindset is not a limitation on progress — it is the condition that allows progress to endure.

CHAPTER IV

Chapter 4: Birth of the ZHA Philosophy

This chapter explores the global challenges caused by uncontrolled technological growth and introduces the foundational thinking behind Sustainable AI with Human-Centric Purpose.

4.1 When Ideas Become Movements

Every transformative philosophy begins with a question.

Not a question asked by institutions, but one asked quietly within an individual observing the world.

Why does progress not always bring peace?
Why do intelligent societies struggle with sustainability?
Why do successful professionals feel disconnected despite achievement?
Why does technological advancement coexist with environmental and social imbalance?

The ZHA Philosophy was born from confronting these questions — not as abstract theory, but as lived experience across professional, social, and community environments.

It did not begin as a grand vision to build a global framework.

It began with observation.

Observation of humanity itself.

4.2 The Personal Journey Behind ZHA

The origins of ZHA emerged at the intersection of multiple worlds:

Enterprise technology transformation, organizational leadership, education and youth development, agriculture and rural ecosystems, social responsibility initiatives.

Working within global technology environments revealed how organizations pursued transformation through processes, tools, and methodologies. Agile frameworks, digital strategies, automation models, and innovation programs promised efficiency and growth.

Yet a recurring realization appeared:

Transformation efforts often failed not because of technology but because of mindset.

Teams resisted change despite modern tools.

Organizations implemented frameworks without cultural alignment.

Innovation initiatives lacked purpose beyond competition.

Simultaneously, engagement with students and communities exposed another reality:

Young generations possessed creativity and passion but lacked structured guidance toward sustainable thinking.

Two worlds enterprise transformation and social development revealed the same insight:

Sustainable change begins with human consciousness.

4.3 From Initiative to Insight

The early activities that later evolved into ZHA were not initially defined as a philosophy.

They appeared as small experiments:

Mentoring students, encouraging community collaboration, introducing sustainability conversations, connecting learning with real-world responsibility.

These initiatives demonstrated something powerful:

When individuals adopted a sustainability mindset, transformation occurred naturally.

- Students became more confident.
- Communities became collaborative.
- Professionals rediscovered meaning in work.

The realization became clear:

Sustainability was not merely an environmental concept.

It was a human operating principle.

4.4 Naming ZHA

The emergence of ZHA represented more than organizational identity.

It symbolized a shift toward holistic thinking.

ZHA came to represent:

Balance between growth and responsibility, alignment between innovation and humanity, integration of knowledge, purpose, and action.

Rather than functioning as a traditional NGO or educational program, ZHA evolved as a living philosophy one that could adapt across sectors, cultures, and generations.

ZHA was never intended to belong to a single domain.

It was designed to bridge many:

- Technology.

- Education.
- Agriculture.
- Society.
- Governance.
- Human development.

4.5 The Core Realization

The defining insight behind ZHA was simple yet transformative:

Human problems cannot be solved by systems created from the same mindset that produced them.

Environmental crises, organizational burnout, social fragmentation, and technological risks shared a common origin, fragmented thinking.

Modern systems often separate:

Economy from ecology, innovation from ethics, intelligence from wisdom, productivity from purpose.

ZHA Philosophy reunites these dimensions.

It views sustainability not as a constraint but as the natural state of balanced civilization.

4.6 From Sustainability to Sustainability Mindset

Traditional sustainability initiatives focus on outcomes:

Reducing emissions, conserving resources, implementing ESG metrics.

While important, outcome-driven approaches struggle when underlying behaviors remain unchanged.

ZHA introduced a deeper approach:

Instead of managing sustainability externally, cultivate it internally.

The Sustainability Mindset framework shifted focus toward:

How individuals think, how leaders decide, how organizations innovate, how communities collaborate.

When mindset changes, sustainable outcomes follow organically.

4.7 The Evolution Toward ZSPC

As the philosophy matured, the need arose to translate ideas into practice.

This led to the development of structured learning ecosystems that would later evolve into the Zha Sustainability Practitioners Community (ZSPC).

ZSPC became a laboratory for testing philosophy through action:

Students practiced sustainability leadership, educators integrated values into learning, communities collaborated on real-world challenges, individuals developed confidence and social responsibility.

The philosophy moved from concept to measurable practice.

Sustainability became experiential rather than theoretical.

4.8 Bridging Technology and Humanity

The next stage of evolution occurred when Artificial Intelligence began reshaping global transformation.

Technology leaders focused on automation, data, and efficiency. Meanwhile, sustainability discussions remained largely separate from technological innovation.

ZHA recognized a critical gap:

AI transformation lacked philosophical grounding.

Organizations were building intelligent systems without defining the human purpose behind intelligence itself.

This realization expanded ZHA from a sustainability initiative into a civilization framework integrating:

Sustainability Mindset, AI Transformation, Product Development, Governance Models, Future Civilization Design.

The philosophy matured into what would become:

Sustainable AI with Human-Centric Purpose.

4.9 ZHA as a Civilization Framework

At this stage, ZHA ceased to be merely an organization.

It became a framework capable of guiding multiple layers of society:

1. Individual Level

   Developing conscious, responsible citizens.
2. Organizational Level

   Transforming enterprises into stewardship-driven institutions.
3. Educational Level

   Preparing future generations for ethical innovation.
4. Technological Level

   Aligning AI development with human and planetary wellbeing.
5. Civilizational Level

   Designing governance and collaboration models for sustainable global progress.

ZHA evolved into a blueprint for conscious civilization.

4.10 The Philosophy of Integration

Many global movements focus on single dimensions:

- Technology movements focus on innovation.
- Environmental movements focus on conservation.
- Economic systems focus on growth.

ZHA Philosophy integrates all dimensions simultaneously.

Its guiding belief:

True sustainability exists when intelligence, economy, society, and nature evolve together.

Integration replaces fragmentation.

Harmony replaces imbalance.

4.11 The Founder's Conviction

The ZHA Philosophy rests upon a personal conviction:

Humanity does not lack intelligence.

Humanity lacks alignment.

When intelligence operates without purpose, disruption emerges.
When intelligence operates with purpose, civilization advances.

Sustainable AI with Human-Centric Purpose represents the culmination of this belief — a commitment to ensure that the most powerful technologies humanity creates remain anchored in compassion, responsibility, and long-term vision.

4.12 From Philosophy to Movement

Every philosophy eventually faces a decision:

Remain an idea, or become a movement.

ZHA chooses movement.

A movement toward:

Conscious leadership, sustainable innovation, collaborative societies, responsible intelligence, and a future civilization built intentionally rather than accidentally.

The journey that began with observation now evolves into global transformation.

Chapter Reflection

ZHA did not begin as a solution to technology.

It began as a response to humanity and in understanding humanity, it discovered the pathway to guiding Artificial Intelligence itself.

PART II — The ZHA Sustainability Mindset

CHAPTER V

Chapter 5: The 12 Sustainability Mindset Principles

This chapter explains how behavioral transformation becomes the first step toward sustainable innovation and ethical technology adoption.

5.1 From Philosophy to Practice

Philosophy inspires transformation, but principles enable action.

The ZHA Sustainability Mindset was not designed as a theoretical model. It emerged from observing how individuals, organizations, and societies struggle to translate sustainability ideals into daily decisions.

Many people support sustainability conceptually, yet real-world behavior often remains unchanged.

Why?

Because sustainability has traditionally been treated as:

- A policy
- A department
- A compliance requirement
- An environmental initiative

ZHA reframes sustainability as a mindset discipline — a way of thinking that guides decisions across technology, leadership, innovation, education, and everyday life.

The twelve Sustainability Mindset Principles form a behavioral operating system enabling Sustainable AI with Human-Centric Purpose.

Each principle addresses a specific imbalance present in modern civilization.

5.2 Principle 1: Feeling Sufficien

Core Idea

True sustainability begins when individuals and institutions understand what is enough.

Modern economies often operate on perpetual dissatisfaction — encouraging continuous consumption, comparison, and accumulation.

Feeling Sufficient does not reject ambition.
It redefines ambition through balance.

Example – Individual Level

A professional constantly upgrades devices despite functional alternatives. Adopting sufficiency thinking shifts focus from ownership to usefulness, reducing waste and psychological dependency on material validation.

Example – Organizational Level

A company avoids unnecessary infrastructure expansion and instead optimizes existing systems using AI-driven efficiency models.

AI Transformation Insight

Sustainable AI designs systems that optimize resource utilization — not endless expansion.

5.3 Principle 2: Encouraging & Rewarding

Core Idea

Human motivation thrives when contributions are recognized meaningfully.

Societies often celebrate achievement only after success becomes visible or profitable.

Sustainability requires recognizing positive contributions early and consistently.

Example – Education

Students contributing to environmental or social initiatives receive recognition through certification and community acknowledgment rather than waiting for academic ranking alone.

Example – Enterprise

Organizations reward collaboration and ethical innovation alongside revenue metrics.

AI Transformation Insight

AI systems should measure and encourage behaviors that improve collective well-being — not only performance efficiency.

5.4 Principle 3: Long-Term Thinking

Core Idea

Civilizations decline when decisions prioritize short-term gain over long-term stability.

Long-Term Thinking introduces generational accountability.

Example – Urban Development

A city planning transportation infrastructure considers environmental impact, public health, and future population needs rather than immediate economic returns.

Example – Technology Product

An AI platform evaluates societal impact before deployment instead of releasing features solely for competitive advantage.

AI Transformation Insight

Product roadmaps must include sustainability metrics extending beyond quarterly performance cycles.

5.5 Principle 4: Risk-Based Decision Making

Core Idea

Innovation without risk awareness becomes disruption.

Risk-based thinking integrates environmental, social, and ethical considerations into decision-making.

Example – Agriculture

Farmers adopting precision agriculture technologies evaluate soil health risks alongside productivity gains.

Example – AI Deployment

Organizations assess algorithmic bias risks before launching automated hiring systems.

AI Transformation Insight

Risk assessment becomes a design requirement rather than a post-implementation audit.

5.6 Principle 5: Gender Equality

Core Idea

Sustainable societies require inclusive participation.

Gender equality is not only a social value — it is an innovation advantage.

Diverse perspectives improve decision quality and reduce systemic bias.

Example – Workplace Innovation

Mixed leadership teams produce more balanced policies addressing employee well-being and productivity.

Example – AI Development

Inclusive datasets and diverse development teams reduce bias in machine learning models.

AI Transformation Insight

Human diversity strengthens artificial intelligence outcomes.

5.7 Principle 6: Sharing & Caring Collaboration

Core Idea

Competition drives advancement, but collaboration sustains civilization.

Modern challenges such as climate change, AI governance, and public health cannot be solved individually.

Example – Community Model

Schools collaborate with local communities to implement sustainability projects instead of operating as isolated institutions.

Example – Industry

Organizations share non-competitive sustainability research to accelerate global progress.

AI Transformation Insight

Collaborative intelligence ecosystems outperform isolated innovation models.

5.8 Principle 7: Environmental Stewardship

Core Idea

Human civilization exists within ecological systems — not outside them.

Environmental stewardship recognizes nature as a partner in progress.

Example – Manufacturing

Factories adopt circular economy models where waste becomes input for new production cycles.

Example – Agriculture + AI

AI monitors soil moisture, biodiversity health, and water usage to prevent ecological degradation.

AI Transformation Insight

AI must optimize for ecological health alongside operational efficiency.

5.9 Principle 8: Fail Fast for Vision

Core Idea

Failure is not the opposite of success; unmanaged failure is.

Fail Fast for Vision encourages experimentation aligned with ethical direction.

Example – Startup Innovation

Teams test sustainable product ideas rapidly while maintaining responsibility toward users and the environment.

Example – Education

Students learn through real-world sustainability projects rather than theoretical examination alone.

AI Transformation Insight

AI innovation should encourage experimentation while maintaining human-centered safeguards.

5.10 Principle 9: Community Engagement

Core Idea

Sustainable solutions cannot be imposed from centralized authority alone.

Communities must participate in decision-making.

Example – Smart Cities

Local citizens contribute data and feedback shaping AI-driven urban planning solutions.

Example – Rural Development

Agricultural technologies are co-designed with farmers rather than introduced top-down.

AI Transformation Insight

Human participation improves trust and adoption of intelligent systems.

5.11 Principle 10: Social Equality

Core Idea

Technology must reduce inequality rather than deepen it.

Social equality ensures access to opportunity, education, and innovation benefits.

Example – Digital Inclusion

AI-enabled education platforms reach underserved populations.

Example – Workforce Transformation

Automation programs include reskilling initiatives rather than workforce displacement.

AI Transformation Insight

AI success should be measured by societal uplift — not exclusive advantage.

5.12 Principle 11: Continuous Learning

Core Idea

Sustainability requires adaptability.

Static societies struggle in dynamic technological environments.

Example – Organizations

Companies continuously update skills through AI learning platforms rather than relying on outdated expertise.

Example – Individuals

Professionals embrace lifelong learning instead of career rigidity.

AI Transformation Insight

AI ecosystems must evolve through continuous feedback and learning loops.

5.13 Principle 12: Resource Conservation

Core Idea

Efficiency is not only economic — it is ethical.

Resource conservation ensures future generations inherit viable ecosystems and economies.

Example – Energy Systems

Data centers adopt renewable energy optimization using AI-based workload management.

Example – Product Design

Companies design products for repairability and reuse rather than planned obsolescence.

AI Transformation Insight

Sustainable AI minimizes energy consumption, computational waste, and environmental footprint.

5.14 The Integrated Power of the Twelve Principles

Each principle alone creates improvement.
Together, they form a Civilization Mindset Framework.

They align:

- Human behavior
- Organizational strategy
- Technological innovation

- Environmental responsibility
- Societal well-being

The principles transform sustainability from aspiration into operational reality.

5.15 Sustainability Mindset as the Foundation for AI Transformation

Before organizations select SDLC methodologies or design AI products, they must establish mindset alignment.

Without these principles:

- Agile becomes accelerated chaos.
- Automation becomes displacement.
- AI becomes amplification of existing imbalance.

With these principles:

- AI becomes regenerative.
- Innovation becomes responsible.
- Technology becomes human-centered.

Sustainable AI is therefore not a technical upgrade. It is a mindset evolution guiding intelligence itself.

Chapter Reflection

Sustainability is not achieved through regulation alone. It is achieved when individuals, organizations, and technologies begin thinking differently. The twelve Sustainability Mindset Principles are not rules. They are the foundation of a conscious civilization.

CHAPTER VI

Chapter 6: Feeling Sufficient vs Over-Consumption

Explains how behavioral transformation becomes the first step toward sustainable innovation and ethical technology adoption.

6.1 The Silent Crisis Behind Modern Progress

Modern civilization celebrates abundance.

Technology promises faster services, unlimited information, instant gratification, and continuous upgrades. Every year brings new devices, smarter platforms, faster algorithms, and more automated systems.

At first glance, this appears to be progress.

Yet beneath technological advancement lies an uncomfortable reality:

- Humanity is experiencing more consumption but less satisfaction.
- People own more tools yet feel more pressured.
- Organizations build more systems yet struggle with sustainability.
- Societies produce more wealth yet face growing inequality and ecological stress.

The problem is not innovation.

The problem is over-consumption driven by psychological scarcity.

This chapter introduces a foundational concept of the ZHA Sustainability Mindset:

- Feeling Sufficient is the starting point of Sustainable AI and ethical innovation.

6.2 Understanding Over-Consumption

Over-consumption is not limited to material goods.

It appears across multiple dimensions:

- **Material Over-Consumption**
  Continuous purchase and disposal of products beyond necessity.
- **Digital Over-Consumption**
  Endless scrolling, data accumulation, attention addiction.
- **Organizational Over-Consumption**
  Building excessive features, tools, reports, and processes.
- **Technological Over-Consumption**
  Deploying AI solutions where human simplicity would suffice.

Modern systems reward expansion rather than balance.

Innovation becomes confused with accumulation.

Example - Personal Technology

A smartphone functioning perfectly is replaced annually not because of necessity but because of perceived obsolescence.

The environmental impact includes:

- mining rare earth materials,
- increased electronic waste,
- energy-intensive manufacturing.

The psychological impact includes:

- comparison culture,
- dependency on novelty,
- reduced appreciation.

Technology advances sustainability declines.

6.3 The Psychology of Scarcity

Over-consumption originates from mindset rather than need.

Human beings operate under an unconscious belief:
"More will make me secure."

Organizations mirror the same belief:

- More automation equals success.
- More data equals intelligence.
- More growth equals stability.

This scarcity mindset drives unsustainable behavior even in resource-rich environments.

Artificial Intelligence, when introduced into such systems, magnifies the problem.

AI optimizes for more because humans define success as more.

Thus, technology amplifies psychological scarcity.

6.4 What Does "Feeling Sufficient" Mean?

Feeling Sufficient does not imply minimalism or rejection of progress.

It represents clarity about enoughness.

It means:

- Using resources consciously.
- Innovating with intention.
- Choosing value over volume.

- Designing technology for wellbeing rather than addiction.

Feeling sufficient shifts the question from:
"How much can we build?"

to

"What truly improves human life?"

Example - Individual Transformation

A professional adopts digital discipline:

- limiting unnecessary notifications,
- choosing tools intentionally,
- focusing on meaningful work rather than constant activity.

Productivity increases while stress decreases.

Sufficiency creates freedom.

6.5 Over-Consumption in Organizations

Organizations frequently equate innovation with expansion.

Common symptoms include:

- Feature overload in products.
- Multiple overlapping software tools.
- Excessive meetings and reporting.
- Technology adoption without strategic purpose.

This leads to innovation fatigue.

- Employees become overwhelmed.

- Customers face complexity.
- Systems become fragile.

Example - Enterprise Software

A company integrates AI into every workflow to appear innovative.

Result:

- employees distrust automation,
- workflows become complicated,
- maintenance costs increase,
- productivity declines.

AI adoption fails not because of technology but because of over-consumption mentality.

6.6 Sustainable Innovation Begins with Behavioral Change

Traditional transformation focuses on technology first.

ZHA proposes the reverse:
Behavioral transformation precedes technological transformation.

Before adopting AI, organizations must ask:

- Do we truly need automation here?
- Does this innovation solve a real problem?
- Will this technology improve human wellbeing?

Sustainable innovation begins when restraint becomes a strategic capability.

Example - Sustainable Product Development

Instead of launching ten features annually, a product team focuses on three high-impact capabilities improving user wellbeing and efficiency.

Outcome:

- higher customer trust,
- reduced maintenance complexity,
- sustainable growth.

Less becomes more.

6.7 Feeling Sufficient and Ethical AI Adoption

Many AI risks originate from over-consumption behaviors:

- collecting excessive personal data,
- optimizing user engagement endlessly,
- automating decisions without necessity.

Feeling sufficient introduces ethical boundaries.

AI systems designed under sufficiency thinking:

- collect only essential data,
- prioritize human agency,
- minimize algorithmic manipulation.

Ethical AI emerges naturally when sufficiency guides design.

Example - Social Media Algorithms

Engagement-maximizing AI prolongs user attention regardless of wellbeing.

A sufficiency-driven platform instead optimizes for:

- meaningful interaction,

- healthy usage duration,
- positive community outcomes.

Technology serves users rather than exploiting attention.

6.8 Environmental Impact of Technological Over-Consumption

Few users recognize the ecological cost of digital systems.

AI infrastructure requires:

- massive data centers,
- high energy consumption,
- cooling systems,
- global hardware supply chains.

Training a large AI model can consume energy comparable to thousands of households.

Sustainable AI therefore requires conscious limitation, not unlimited scaling.

Example - Green AI Practices

Organizations adopt:

- energy-efficient algorithms,
- shared model architectures,
- renewable-powered data centers.

Innovation continues while environmental impact reduces.

Sufficiency aligns intelligence with planetary limits.

6.9 Sufficiency Thinking in SDLC and Product Development

Feeling sufficient fundamentally changes how organizations choose SDLC methodologies.

Traditional SDLC goals:

- faster delivery,
- maximum feature output,
- continuous expansion.

Sustainable SDLC goals:

- meaningful delivery,
- maintainable systems,
- long-term product resilience.

Agile Under Sufficiency Mindset

Agile becomes:

- user-value driven,
- sustainability-aware,
- outcome-focused rather than velocity-obsessed.

Teams measure success by impact, not story points.

DevOps Under Sufficiency Mindset

Automation focuses on:

- reliability,
- energy efficiency,
- reduced technical debt.

Technology serves stability instead of endless acceleration.

6.10 Sufficiency as Leadership Maturity

Leaders operating from sufficiency thinking demonstrate:

- calm decision-making,
- long-term vision,
- ethical clarity,
- reduced reaction to competitive pressure.

They recognize that sustainable success comes from alignment rather than expansion.

In AI transformation, leadership mindset determines whether technology becomes regenerative or disruptive.

6.11 The Civilization Perspective

Historically, civilizations collapsed when consumption exceeded regeneration.

The AI era introduces a new dimension:
Over-consumption of intelligence itself.

When:

- automation replaces meaning,
- data replaces wisdom,
- growth replaces balance,

Civilization risks losing its human center.

Feeling sufficient restores equilibrium between intelligence and humanity.

6.12 Behavioral Transformation: The First Step Toward Sustainable AI

Sustainable AI adoption begins not in engineering departments but in human awareness.

Behavioral transformation includes:

- redefining success,
- embracing mindful innovation,
- prioritizing wellbeing over expansion,
- aligning technology with purpose.

Only then can AI transformation become sustainable.

Without sufficiency, innovation accelerates disruption. With sufficiency, innovation enables civilization evolution.

Chapter Reflection

Sustainability does not begin with technology. It begins when humans realize they already have enough to build wisely. Feeling sufficient transforms innovation from consumption into contribution.

CHAPTER VII

Chapter 7: Social Responsibility as Civilization Foundation

Explains how behavioral transformation becomes the first step toward sustainable innovation and ethical technology adoption.

7.1 Civilization Is Built on Responsibility

Civilizations are not defined by their technology, wealth, or military power. They are defined by how responsibly they treat:

- People
- Knowledge
- Resources
- Future generations

History shows that societies flourish when responsibility becomes a collective culture rather than enforced regulation.

Ancient civilizations built irrigation systems, knowledge institutions, and community governance structures because individuals understood their role within a larger social ecosystem.

Modern society, however, faces a paradox:

Human capability has expanded globally, yet the sense of shared responsibility has weakened.

Technology connects humanity digitally, but responsibility often remains fragmented.

The AI era demands the rediscovery of social responsibility as the foundation of civilization stability.

7.2 From Individual Success to Collective Wellbeing

Modern economic systems reward individual achievement:

- Career success
- Financial growth
- Innovation leadership
- Competitive advantage

These achievements drive progress but can unintentionally shift focus away from collective wellbeing.

Social responsibility reframes success:

Success is sustainable only when it benefits society alongside the individual.

This principle does not diminish ambition; it elevates it.

True leadership emerges when personal success contributes to societal progress.

Example – Entrepreneurial Responsibility

A startup developing AI analytics tools considers not only profitability but also:

Workforce reskilling impacts, Ethical data usage, Accessibility for smaller communities

The company grows commercially while strengthening social trust.

Responsibility becomes a competitive advantage.

7.3 The Responsibility Gap in the AI Era

Artificial Intelligence introduces unprecedented decision-making power.

Algorithms influence:

- Hiring decisions
- Credit approvals
- Healthcare diagnostics
- Education pathways
- Public discourse

Yet responsibility for these outcomes often becomes diffused.

Engineers build systems.

Executives deploy them.

Users interact with them.

Regulators react afterward.

Who, then, is responsible?

The AI era exposes what can be called the Responsibility Gap, where technology advances faster than collective accountability structures.

Closing this gap requires embedding responsibility into mindset, culture, and governance from the beginning.

7.4 Social Responsibility Beyond Compliance

Organizations frequently interpret responsibility as compliance:

- ESG reporting
- Regulatory adherence
- Ethical guidelines
- Corporate social responsibility programs

While necessary, compliance alone cannot create sustainable civilization.

Compliance asks:
"What must we avoid doing wrong?"

Social responsibility asks:
"What good must we intentionally create?"

The difference is transformational.

Responsibility shifts from obligation to purpose.

Example – Corporate Transformation

Instead of sponsoring isolated CSR events, an organization integrates sustainability into:

- Product design
- Employee development
- Supply chain decisions
- Innovation strategy

Responsibility becomes operational rather than symbolic.

7.5 Social Responsibility and Sustainable Innovation

Innovation detached from responsibility often generates unintended harm.

Examples across history include:

Industrial pollution following manufacturing expansion, Social media algorithms amplifying division, Automation creating workforce displacement without transition planning, Sustainable innovation requires anticipatory responsibility.

Innovators must consider consequences before scale.

Example – AI Healthcare System

An AI diagnostic tool is introduced with:

- Human oversight mechanisms
- Bias monitoring
- Accessibility across socioeconomic groups
- Innovation improves healthcare outcomes while maintaining trust.
- Responsible design prevents future crisis.

7.6 The Role of Communities

Civilization is sustained not by institutions alone but by communities.

Communities provide:

- Local wisdom
- Cultural stability
- Human connection
- Participatory governance

Technological transformation often occurs at national or corporate levels, yet its impact is experienced locally.

Social responsibility therefore includes community engagement as a core design principle.

Example – Smart City Development

Instead of imposing AI surveillance systems unilaterally, city planners engage residents to define acceptable use, privacy expectations, and social benefits.

Technology adoption becomes collaborative rather than imposed.

Trust increases.

7.7 Education as Responsibility Infrastructure

Education represents civilization's most powerful responsibility mechanism.

What societies teach their youth determines future technological behavior.

Traditional education focuses heavily on skills acquisition.

The ZHA framework introduces an expanded purpose:

Education must cultivate responsible citizens capable of guiding intelligent technologies ethically.

Students must learn:

- Sustainability thinking
- Empathy
- Collaborative leadership
- Systems awareness
- Ethical innovation

Education becomes preparation for stewardship rather than employment alone.

7.8 Social Responsibility in Organizations

Organizations operating under social responsibility principles demonstrate distinctive characteristics:

- Purpose-Driven Leadership

  Leaders consider societal impact alongside financial outcomes.

- Inclusive Innovation

Products are designed to benefit diverse populations.

- Workforce Care

  Automation includes reskilling and human development programs.

- Environmental Consciousness

  Operational efficiency aligns with ecological sustainability.

Such organizations evolve from profit engines into civilization partners.

7.9 Responsibility and AI Transformation Leadership

AI transformation leaders face a critical choice.

They can pursue:

Efficiency-first transformation, Responsibility-led transformation, Responsibility-led AI transformation asks:

- Does automation enhance human dignity?
- Does data usage respect autonomy?
- Does innovation reduce inequality?
- Does technology strengthen society?

When responsibility becomes leadership philosophy, AI transformation becomes sustainable.

7.10 ZHA Sustainability Mindset and Social Responsibility

Within the ZHA framework, social responsibility emerges naturally from the twelve sustainability mindset principles.

Feeling sufficient reduces exploitation.

Collaboration strengthens communities.

Long-term thinking protects future generations.

Resource conservation safeguards ecosystems.

Responsibility is no longer external enforcement — it becomes internal culture.

This transformation represents the transition from reactive governance to conscious civilization.

7.11 The Civilization Perspective

Civilizations do not collapse due to lack of intelligence.

They decline when responsibility erodes.

The AI age magnifies both potential and risk:

Intelligence without responsibility accelerates instability.

Intelligence guided by responsibility enables flourishing societies.

Social responsibility becomes the stabilizing force allowing technological advancement to coexist with human wellbeing.

7.12 Toward a Conscious Civilization

A conscious civilization recognizes interconnectedness:

- Individual actions affect communities.
- Corporate decisions affect ecosystems.
- Technological innovations affect future generations.
- Social responsibility transforms this awareness into action.

It ensures that progress remains aligned with humanity's highest values.

The ZHA Civilization Framework positions responsibility not as a burden but as humanity's greatest evolutionary achievement.

Chapter Reflection

Technology may shape civilization's tools.
Responsibility shapes civilization's destiny.

When societies embrace responsibility collectively, innovation becomes sustainable, intelligence becomes humane, and civilization becomes conscious.

CHAPTER VIII

Chapter 8: Long-Term Thinking and Planet Stewardship

Explains how behavioral transformation becomes the first step toward sustainable innovation, ethical technology adoption, and aligning AI transformation with the future of humanity.

8.1 The Short-Term Civilization Problem

Modern civilization operates under an invisible but powerful assumption:

Success must be immediate.

- Quarterly profits determine strategy.
- Product releases define innovation.
- Technology adoption follows market urgency.

This short-term orientation has enabled rapid growth but introduced deep systemic risks:

- Environmental degradation
- Technology misuse
- Organizational instability
- Social fragmentation

Artificial Intelligence intensifies this challenge because AI accelerates decision-making faster than human reflection.

When intelligence scales rapidly under short-term thinking, civilization risks optimizing itself toward long-term harm.

The fundamental challenge of the AI era therefore becomes:

Can humanity think long-term while operating in a short-term world?

8.2 What Is Long-Term Thinking?

Long-term thinking is not prediction; it is responsibility extended across generations.

It asks leaders to consider:

Consequences beyond immediate success, Impacts beyond organizational boundaries, Outcomes beyond current lifetimes, Long-term thinking recognizes that civilization is an inheritance shared across generations.

Every technological decision today shapes the world future generations will inhabit.

Example – Infrastructure Planning

A city builds transportation systems designed for 50 years rather than 5 years.

Benefits include:

- Reduced environmental impact
- Economic resilience
- Improved quality of life
- Short-term cost increases create long-term stability.

8.3 Planet Stewardship: Humanity's New Role

Historically, humanity viewed nature as a resource.

Industrial development reinforced extraction:

- Forests became timber
- Rivers became energy sources
- Land became production assets

The AI era demands a new perspective:

Humanity must evolve from resource consumer to planetary steward.

Planet stewardship recognizes Earth as a living system supporting civilization itself.

Technology must now operate within ecological boundaries rather than attempting to dominate them.

Stewardship Mindset

Stewardship means:

- Protecting biodiversity
- Regenerating ecosystems
- Balancing consumption

Designing technologies compatible with nature

AI becomes a tool for stewardship rather than exploitation.

8.4 Why AI Transformation Requires Long-Term Thinking

AI transformation initiatives often focus on immediate business outcomes:

- Automation efficiency
- Operational savings
- Competitive advantage
- Rapid innovation

While valuable, these goals ignore broader implications:

Energy consumption of large models, Workforce displacement, Algorithmic bias persistence, Social dependence on automation, Without long-term thinking, AI transformation risks becoming technologically successful but civilizationally harmful.

Example – AI Model Training

Training large AI models requires massive computational power and energy.

Short-term thinking:
"Build larger models for performance."

Long-term thinking:
"How can intelligent systems deliver value while minimizing environmental footprint?"

Organizations adopting sustainable AI invest in:

- Efficient model architectures
- Shared training infrastructure
- Renewable energy-powered data centers
- Innovation continues without ecological sacrifice.

8.5 The Time Horizon of Leaders

Leadership maturity can be measured by time horizon.

Short-term leaders optimize for immediate gains.
Long-term leaders optimize for sustainable impact.

AI transformation leaders must operate at the civilizational level.

They are not merely deploying technology; they are shaping humanity's future relationship with intelligence.

The ZHA framework elevates AI leadership from operational execution to planetary responsibility.

8.6 Long-Term Thinking in Product Development

Traditional product development prioritizes:

- Rapid feature release
- Market capture
- Short lifecycle innovation

This creates:

- Software waste
- Technical debt
- Unstable ecosystems
- User fatigue

Sustainable AI product development introduces longevity design.

Example – Sustainable Digital Platform

Instead of releasing continuous disruptive updates, a platform designs features that:

- Remain usable for years
- Respect user attention
- Minimize dependency cycles

Products become stable digital environments rather than addictive consumption engines.

8.7 Planet Stewardship Through AI

Artificial Intelligence uniquely enables humanity to protect the planet at unprecedented scale.

Environmental MonitoringAI analyzes satellite imagery to detect deforestation and ecosystem damage in real time.

- Climate Modeling

Machine learning predicts climate patterns, enabling proactive mitigation strategies.

- Precision Agriculture

AI optimizes water usage, fertilizer distribution, and soil health management.

- Energy Optimization

Smart grids powered by AI reduce energy waste and integrate renewable resources efficiently.

AI becomes humanity's planetary nervous system — sensing, learning, and protecting ecological balance.

8.8 Organizational Transformation Toward Stewardship

Organizations embracing long-term thinking undergo fundamental transformation.

They evolve from:

Profit Maximizers → Value Stewards

Characteristics include:

- Sustainability integrated into core strategy
- Innovation aligned with societal impact
- Leadership accountability beyond shareholders
- Resilience prioritized over rapid expansion

Such organizations attract trust, talent, and longevity.

Example – Sustainable Enterprise AI Strategy

An enterprise implementing AI automation also invests in:

- Employee reskilling programs
- Ethical AI governance boards
- Environmental impact measurement
- Transformation becomes inclusive rather than disruptive.

8.9 Long-Term Thinking and SDLC Selection

The ZHA framework introduces a strategic shift:

SDLC methodology must align with long-term sustainability goals.

Instead of asking:
"What delivers fastest?"

Organizations ask:
"What sustains value longest?"

Agile with Long-Term Thinking

Focus on sustainable iteration

Avoid feature inflation

Maintain system durability and maintainability

DevOps with Stewardship

Energy-efficient infrastructure

Responsible automation

Reduced technical waste

AI Lifecycle Governance

Continuous monitoring of societal impact

Lifecycle sustainability assessment

Methodology becomes a reflection of values.

8.10 The Intergenerational Responsibility of AI

Artificial Intelligence represents the first technology capable of influencing multiple future generations simultaneously.

Decisions made today may persist for decades through:

- Automated governance systems
- Intelligent infrastructure
- Learning algorithms shaping knowledge distribution

Thus, AI development carries intergenerational responsibility.

Future generations will not judge humanity by technological sophistication alone.

They will judge whether intelligence was used wisely.

8.11 The ZHA Perspective: From Growth to Regeneration

Long-term thinking transforms innovation philosophy:

- Growth extracts.
- Regeneration restores.

Sustainable AI enables regenerative systems where technology:

- Rebuilds ecosystems
- Strengthens communities
- Enhances human wellbeing
- Preserves planetary balance

Civilization transitions from consumption-driven progress to regeneration-driven evolution.

8.12 Toward a Conscious Planetary Civilization

Planet stewardship represents humanity's maturation as a species.

For the first time, humans possess the intelligence to understand planetary systems and the technology to influence them globally.

The question is no longer whether we can shape the planet.

The question is whether we will do so responsibly.

Sustainable AI with human-centric purpose becomes the bridge between intelligence and stewardship.

Through long-term thinking, AI transformation evolves into civilization transformation.

Chapter Reflection

Short-term intelligence builds powerful systems.
Long-term wisdom sustains civilization.

AI transformation guided by planetary stewardship ensures that progress does not compromise the future it seeks to create.

PART III - AI Transformation Using ZHA Framework

CHAPTER IX

Chapter 9: Human-Centric AI Transformation

Describes organizational transformation practices aligning Artificial Intelligence with ESG principles, sustainability values, and long-term societal impact.

9.1 The Transformation Humanity Is Facing

Artificial Intelligence is often described as a technological revolution.

In reality, AI represents something deeper:

a transformation of how humanity thinks, works, decides, and evolves.

- Previous technological revolutions enhanced human physical capability.
- Machines amplified strength.
- Computers amplified calculation.
- The internet amplified communication.

Artificial Intelligence amplifies decision-making itself.

For the first time, humans are delegating elements of cognition to machines.

This shift requires a new transformation model — one that places humans at the center rather than at the margins of innovation.

This is the foundation of Human-Centric AI Transformation.

9.2 The Problem with Technology-Centric Transformation

Most organizations begin AI transformation incorrectly.

Typical starting questions include:

Which AI tools should we adopt?

How fast can we automate?

How much cost reduction is possible?

How can we outperform competitors?

Technology becomes the focal point.

However, technology-centric transformation produces predictable challenges:

- Employee resistance
- Ethical concerns
- Loss of trust
- Unintended societal impact
- Unsustainable innovation cycles

The failure lies not in AI capability but in transformation philosophy.

AI adoption must begin with human purpose, not technological possibility.

9.3 Defining Human-Centric AI

Human-Centric AI Transformation ensures that intelligent systems:

- Enhance human capability
- Preserve human dignity
- Strengthen communities
- Support planetary sustainability
- AI becomes a collaborator rather than a replacement.

Human-centric transformation asks:

How does AI empower people?

Does automation create meaning or remove it?

Are humans becoming wiser alongside smarter machines?

Technology serves humanity — not the reverse.

9.4 The ZHA AI Transformation Model

The ZHA Framework introduces a structured evolution path for AI transformation:

1. ALIGN — Purpose Before Technology

Most organizations begin transformation by purchasing AI tools.
ZHA begins transformation by asking:

Why should intelligence be introduced at all?

Alignment ensures AI serves human and societal goals rather than technological excitement.

Without alignment:

Automation amplifies inefficiency, AI accelerates poor decisions, Innovation creates unintended harm, Key Alignment Activities, Define organizational purpose, Identify who benefits, Establish sustainability principles, Assess cultural and ethical readiness

Example – Bank AI Transformation

Misaligned approach:
A bank introduces AI credit scoring purely for efficiency.
Result: bias complaints, regulatory risks, customer distrust.

Aligned approach:
The bank defines its purpose — expand financial inclusion, improve fairness, support underserved communities.

AI design follows purpose.
Outcome: higher trust and sustainable innovation.

SDLC begins with values, not code.

2. AUTOMATE — Responsible Intelligence Deployment

Automation is not the goal.
Automation should eliminate friction, not humanity.

Responsible automation asks:

Does this free humans for meaningful work?

Does it reduce stress?

Does it improve quality of life?

Principles

Automate repetitive tasks

Preserve human judgment

Maintain accountability

Avoid blind automation

Example – Healthcare

Poor automation: fully automated diagnosis removes physician interaction.
Result: loss of trust and empathy.

Responsible automation: AI analyzes scans; doctors focus on patient care and complex decisions.
Automation enhances humanity.

Employees shift from task execution to value creation.

3. AUGMENT — Expanding Human Capability

The greatest impact of AI is augmentation, not automation.

AI becomes a cognitive partner helping humans:

- Think better
- Learn faster
- Decide wisely
- Create innovatively

Example – Education

AI enables personalized learning paths and adaptive assessments.
Teachers focus on mentorship and human development.

Example – Agriculture

AI analyzes soil health, weather patterns, and crop disease risk.
Farmers make better decisions with intelligence support.

Technology strengthens human wisdom.

4. INNOVATE — Human-Centered Creation

Only after alignment, automation, and augmentation should organizations pursue disruptive innovation.

Innovation without maturity creates chaos.
Human-centered innovation solves real problems and enhances societal wellbeing.

Example – Smart Cities

Technology-driven model: mass surveillance infrastructure.

Human-centered model: AI-enabled mobility, energy efficiency, participatory governance systems.

Innovation becomes civilization building.

5. SUSTAIN & SCALE — From Innovation to Civilization Impact

Many organizations innovate successfully but fail to sustain impact.

Scaling without sustainability causes:

- Ethical crises
- Environmental strain
- Social backlash

Sustainable scaling ensures AI continues benefiting humanity over decades.

Sustainability Dimensions

- Ethical sustainability
- Environmental sustainability
- Social sustainability
- Organizational sustainability

AI evolves from product to trusted public infrastructure.

9.5 Alignment Before Automation

Many organizations automate inefficient processes without questioning necessity.

Human-centric transformation begins with alignment.

Example – Customer Service

Technology-centric: deploy chatbots to reduce workforce costs.

Human-centric: automate repetitive tasks while enabling human agents to handle complex, empathetic interactions.

Automation enhances humanity rather than replacing it.

9.6 AI as Human Augmentation

AI can:

- Assist doctors in diagnosis
- Support teachers with personalized learning
- Help engineers analyze complex systems
- Enable farmers to predict crop health

Example – Healthcare

AI analyzes medical imaging rapidly, reducing errors and enabling early detection.
Physicians spend more time with patients.

Technology restores human connection.

9.7 Workforce Transformation: Fear vs Opportunity

Human-Centric AI reframes job displacement.

The future of work is not about fewer humans.
It is about different humans.

Organizations must transition from workforce reduction to workforce evolution.

Example – Manufacturing

Instead of eliminating workers, companies retrain employees as:

- Automation supervisors

- Data analysts
- Maintenance specialists
- Innovation contributors
- AI becomes a career multiplier.

9.8 Ethical Intelligence in Decision Systems

AI influences hiring, credit approval, healthcare prioritization, and education assessment.

Human-centric transformation ensures:

- Transparency
- Human oversight
- Bias monitoring

Example – AI Hiring

AI provides explainable recommendations.
Diversity checkpoints are built in.
Final decisions remain human.

AI informs decisions — humans remain accountable.

9.9 Human-Centric Product Development

Traditional product development optimizes engagement and revenue.

Human-centric development optimizes wellbeing and trust.

Example – Digital Platforms

Technology-centric model: maximize screen time.

Human-centric model: encourage healthy usage habits.

Long-term trust becomes competitive advantage.

9.10 Sustainable AI Infrastructure

Human-centric transformation includes infrastructure choices.

Organizations must consider:

- Energy consumption
- Carbon footprint
- Hardware lifecycle
- Responsible data storage

Example – Green Data Centers

AI workloads scheduled according to renewable energy availability reduce environmental impact without sacrificing performance.

Technology aligns with planetary stewardship.

9.11 Leadership in Human-Centric Transformation

AI transformation leaders evolve beyond technical management.

They become:

- Ethical navigators
- Cultural architects
- Sustainability stewards
- Civilization designers

Leadership expands from operational efficiency to societal impact.

Human-centric leaders ask:

Will this technology improve human life?

Does automation preserve dignity?

Are we creating dependence or empowerment?

What legacy will this system leave?

Leadership becomes conscious.

9.12 Organizational Culture Shift

Human-centric AI requires cultural transformation.

Organizations must move from:

- Efficiency obsession → Purpose orientation
- Competition-only mindset → Collaborative value creation
- Short-term gains → Long-term stewardship

Culture becomes the true transformation engine.

9.13 Measuring Human-Centric AI Success

Success metrics must evolve beyond financial indicators.

Human-centric AI evaluates:

- Employee wellbeing
- Social impact
- Environmental sustainability
- Trust levels
- Long-term resilience

Innovation success becomes multidimensional.

9.14 AI Transformation as Civilization Evolution

Human-Centric AI Transformation represents more than organizational change.

It signals humanity's transition toward a conscious technological civilization.

The defining question becomes:

Will AI make humanity more efficient — or more human?

The ZHA framework asserts:

Technology achieves its highest purpose when it strengthens compassion, creativity, collaboration, and stewardship.

Chapter Reflection

Artificial Intelligence does not diminish humanity.
It challenges humanity to evolve.

Human-Centric AI Transformation ensures that intelligence advances alongside wisdom.

CHAPTER X

AI-Infused Product Development Lifecycle

Describes organizational transformation practices aligning Artificial Intelligence with ESG principles, sustainability values, and long-term societal impact.

10.1 The Evolution of Product Development

Product development has evolved through multiple technological eras.

Today's products are no longer static software applications.

They are:

- Adaptive
- Learning
- Predictive
- Continuously evolving systems

Traditional Software Development Life Cycle (SDLC) models were designed for deterministic systems — systems behaving exactly as programmed.

Artificial Intelligence introduces probabilistic systems that learn and evolve after deployment.

Therefore, product development must evolve.

The ZHA framework introduces the AI-Infused Product Development Lifecycle, integrating:

- Human-Centric Purpose
- Sustainability Mindset
- Responsible AI Governance

- SDLC Best Practices
- Global Technology Standards

10.2 Why Traditional SDLC Alone Is Not Enough

Traditional SDLC focuses on:

- Requirements
- Development
- Testing
- Deployment
- Maintenance

However, AI systems introduce new realities:

- Models learn continuously
- Data changes behavior
- Ethical risks evolve over time
- Environmental costs increase with scale

Thus, AI products require lifecycle thinking beyond engineering execution.

AI development must integrate:

- Purpose
- Ethics
- Sustainability
- Governance
- Continuous learning

10.3 The ZHA AI-Infused Lifecycle

The ZHA AI Product Lifecycle aligns with the Human-Centric AI Transformation model:

- Align
- Design
- Build
- Validate
- Deploy
- Augment & Learn
- Sustain & Govern

This lifecycle works across all SDLC methodologies.

Stage 1 — ALIGN

Purpose & Sustainability Definition

Before writing code, organizations define:

- Human value proposition
- Ethical boundaries
- ESG considerations
- Social impact expectations

Example — AI Recruitment Platform

Instead of starting with model development, the alignment phase defines:

- Fairness goals
- Diversity inclusion metrics
- Bias mitigation requirements
- Alignment becomes part of requirements engineering.

SDLC begins with values.

Stage 2 — DESIGN

Human-Centric Architecture

AI design includes:

- Data governance
- Explainability mechanisms
- Sustainability architecture
- Human oversight systems

Example — Healthcare AI

Design integrates:

- Clinician override capability
- Transparent decision explanation
- Patient data privacy safeguards
- Architecture embeds ethics by design.

Stage 3 — BUILD

Responsible Development

Development teams build:

- Models
- Pipelines
- Automation workflows
- User interfaces

ZHA adds:

- Sustainable coding practices
- Optimized model training
- Energy-aware infrastructure

Example — AI Model Development

Instead of repeatedly training oversized models, teams:

- Reuse pretrained models
- Reduce computational waste
- Track carbon footprint
- Engineering becomes environmentally conscious.

Stage 4 — VALIDATE

Ethical, Technical & Sustainability Testing

AI validation extends beyond functional testing.

Testing includes:

- Bias evaluation
- Explainability verification
- Environmental impact assessment
- Human usability testing

Example — Financial AI

Validation ensures:

- No demographic bias
- Regulatory compliance
- Transparent decision logic
- Quality expands from performance to responsibility.

Stage 5 — DEPLOY

Responsible Release

Deployment considers:

- Operational stability
- Societal readiness
- User education
- Governance monitoring

Example — Smart City AI

Instead of full-scale rollout:

Pilot zones are launched, Community feedback is gathered, Systems are refined collaboratively, Deployment becomes participatory.

Stage 6 — AUGMENT & LEARN

Continuous Intelligence Evolution

Unlike traditional software, AI improves post-deployment.

Organizations must:

- Monitor performance
- Retrain models responsibly
- Integrate human feedback loops

Example — Education AI Platform

The system adapts to student progress while teachers provide contextual feedback.

AI evolves alongside human learning.

Stage 7 — SUSTAIN & GOVERN

Long-Term Civilization Responsibility

The final stage ensures AI remains beneficial over decades.

Governance includes:

- Ethical audits
- Sustainability reporting
- Bias monitoring
- Energy optimization
- Societal impact evaluation

Example — Global AI Platform

The organization establishes:

- A Responsible AI board
- ESG monitoring dashboards
- Sustainability KPIs
- Innovation becomes accountable infrastructure.

10.4 Mapping ZHA Lifecycle to SDLC Methodologies

The ZHA lifecycle enhances existing SDLC models:

- Agile

  Adds ethical backlog items

  Introduces sustainability acceptance criteria
- Waterfall

  Integrates governance checkpoints

  Supports regulated AI environments
- DevOps / MLOps

  Ensures continuous monitoring

Embeds responsible release gates

- Scaled Agile (SAFe)

  Aligns portfolio strategy with sustainability goals

- CMMI & Quality Standards

  Expands maturity beyond process quality to civilization responsibility

  Methodology becomes an expression of values.

10.5 AI Product Development Maturity Model

AI maturity reflects mindset evolution as much as technical capability.

The ZHA framework defines five levels:

Technology Use → Human Empowerment → Civilization Responsibility

Level 1 — Tool Adoption

AI as Experiment

Isolated tool purchases

No governance framework

Limited leadership understanding

Example: A chatbot answers basic FAQs but fails complex interactions.
AI exists, transformation does not.

Level 2 — Process Automation

AI Improves Efficiency

Workflow automation

Predictive analytics

Cost optimization

Example: Insurance claim automation reduces processing time but increases workforce anxiety.

Efficiency improves; mindset does not.

Level 3 — Human Augmentation

AI as Collaborative Intelligence

Human-in-the-loop systems

Reskilling programs
Augmented workflows

Example: Healthcare AI supports doctors in diagnosis while preserving human judgment.

AI becomes cognitive assistant.

Level 4 — Sustainable Innovation

AI Creates Responsible Value

ESG integrated into strategy

Ethical AI governance boards

Sustainable infrastructure choices

Example: Smart agriculture platform improves yield while conserving water and biodiversity.

Innovation becomes regenerative.

Level 5 — Civilization Stewardship

AI Serves Humanity and the Planet

At this level:

AI becomes societal infrastructure

Governance is human-centric

Planet stewardship is embedded

Long-term civilization impact is prioritized

Example: A global AI ecosystem integrating education, ESG metrics, carbon analytics, and ethical governance systems.

Engineering becomes stewardship.

Evolution Across Levels

Most organizations remain at Level 2 or 3.

Few reach Level 4.

Very few achieve Level 5.

The future leaders of the AI era will move beyond automation and innovation into civilization stewardship.

ZHA defines the pathway.

Chapter Reflection

Artificial Intelligence maturity is not measured by computational power.

It is measured by how responsibly intelligence shapes humanity's future.

The highest form of innovation is stewardship.

10.6 Sustainable AI as Product Philosophy

Products are no longer software artifacts.
They become living ecosystems affecting:

- Economies
- Behaviors
- Environments
- Societies

AI product leaders therefore become civilization designers.

10.6.1 The End of the Traditional Product Era

For decades, products were understood as artifacts:

- A physical object
- A software application
- A released version
- A deliverable owned by an organization

Traditional development followed a predictable lifecycle:

- Design → Build → Release → Maintain
- Once delivered, behavior remained largely static.
- Artificial Intelligence fundamentally changes this assumption.

AI-powered products:

- Learn
- Evolve
- Influence

- Reshape human systems continuously

A modern AI product does not merely perform tasks.
It participates in society.

10.6.2 Products as Living Ecosystems

AI products function more like ecosystems than tools.

They continuously interact with:

- Users
- Data
- Environments
- Institutions
- Cultures

Every interaction feeds back into the system, influencing future outcomes.

Thus, AI products become living ecosystems affecting multiple dimensions of civilization.

Example — Navigation Platforms

A navigation app does more than show routes. It influences:

- Traffic distribution
- Fuel consumption
- Urban planning patterns
- Local business economics
- Environmental emissions

A single product reshapes entire cities.
The product becomes infrastructure.

10.6.3 Impact on Economies

AI products increasingly shape economic behavior.

Recommendation engines influence purchasing decisions, Algorithmic trading shapes financial markets, Digital platforms redefine employment structures

Example — E-Commerce Recommendation AI

Product recommendations:

- Determine which businesses grow
- Influence consumer demand
- Shape supply chains globally

The algorithm indirectly influences economic opportunity.

Product leaders are therefore shaping economic ecosystems.
This responsibility transcends software engineering.

10.6.4 Impact on Human Behavior

AI products influence how humans think, communicate, and behave.

Unlike traditional tools, AI adapts to human psychology.

Example — Social Media Algorithms

Engagement-driven AI systems can unintentionally:

- Amplify emotional reactions
- Reinforce cognitive bias
- Encourage addictive behavior patterns

The product is no longer neutral.

It becomes a behavioral architect.

Human attention itself becomes shaped by product design decisions.

Sustainable Product Perspective

A sustainable AI product asks:

- Does this system improve human wellbeing?
- Does it encourage healthy engagement?
- Does it strengthen social connection?

Product philosophy shifts from attention capture to human flourishing.

10.6.5 Impact on Environments

Digital products appear intangible, yet AI infrastructure consumes significant planetary resources.

AI requires:

- Large-scale data centers
- Electricity
- Cooling systems
- Hardware manufacturing
- Rare earth minerals

Every algorithm carries an environmental footprint.

Example — Large Language Models

Training large-scale AI models can consume energy comparable to thousands of homes.

A sustainable product leader considers:

- Model efficiency
- Energy sources
- Lifecycle environmental impact
- AI architecture becomes an environmental decision.

10.6.6 Impact on Societies

AI products influence governance, education, healthcare, and democracy.

They determine:

- Access to knowledge
- Employment opportunities
- Civic participation
- Information credibility

Example — Educational AI Platforms

AI tutoring systems influence:

- How students learn
- What knowledge is prioritized
- Future workforce skills

The product shapes an entire generation's learning patterns.

Product development becomes societal design.

10.6.7 The Rise of the Civilization Designer

In the AI era, product leaders are no longer merely:

- Product Managers
- Engineers
- Architects

- Designers

They become Civilization Designers.

Every product decision answers civilizational questions:

- What behaviors does this system encourage?
- Who benefits?
- Who may be excluded?
- What long-term societal structures emerge?

Product leadership evolves into stewardship.

New Responsibilities of AI Product Leaders

AI product leaders must integrate:

- Technology vision
- Ethical reasoning
- Sustainability awareness
- Social impact understanding
- Long-term governance thinking

The role expands from delivery management to future design.

10.6.8 Sustainable AI Product Philosophy

The ZHA framework introduces Sustainable AI as a guiding product philosophy.

Instead of asking:

"How do we build successful products?"

We ask:

"How do we build products that sustain civilization?"

Sustainable AI products demonstrate:

- Human-Centric Design
  Enhancing dignity, creativity, and wellbeing.
- Ecological Awareness
  Minimizing environmental footprint.
- Social Responsibility
  Reducing inequality and promoting inclusion.
- Long-Term Stability
  Remaining beneficial across generations.

10.6.9 Product Philosophy Across SDLC Methodologies

Sustainable AI integrates into existing SDLC models.

Agile

- Traditional Agile:
  Deliver features rapidly.
- Sustainable Agile:
  Deliver meaningful value sustainably.

Backlogs include:

- Ethical considerations
- Sustainability criteria
- Long-term maintainability
- DevOps / MLOps

Traditional DevOps:

Speed and automation.

Sustainable DevOps:

Responsible automation with environmental awareness.

Includes:

- Energy-aware deployments
- Ethical monitoring pipelines
- Waterfall / Regulated Development

Sustainable AI enhances:

- Governance checkpoints
- Societal impact reviews
- Compliance alignment

Release becomes responsibility, not completion.

10.6.10 From Products to Platforms to Ecosystems

The evolution of digital systems follows:
 Products → Platforms → Ecosystems

ZHA defines the next evolution:

Civilization Infrastructure

AI products become foundational systems supporting humanity itself.

10.6.11 The Ethical Advantage of Sustainable Products

Organizations adopting Sustainable AI philosophy gain strategic advantage:

- Higher user trust
- Regulatory resilience

- Long-term market relevance
- Stronger brand legitimacy
- Reduced societal backlash
- Ethics and sustainability become innovation accelerators.

10.6.12 ZHA Insight: Technology as Living Responsibility

The greatest misconception of the digital age is believing products end at deployment.

In reality, AI products continue shaping civilization long after release:

- Every update changes behavior
- Every model retraining alters outcomes
- Every algorithm influences society
- Product development becomes an ongoing moral responsibility.

ZHA positions AI creators not as software builders but as caretakers of intelligent ecosystems.

Chapter Reflection

Products are no longer lines of code.
They are living forces shaping economies, environments, and societies.

The future belongs to organizations that recognize AI product development as an act of civilization stewardship.

10.7 The ZHA Insight

The future of product development is not faster delivery.
It is responsible intelligence creation.

AI-Infused Product Development ensures that:

- Innovation remains human-centered

- Technology remains sustainable
- Intelligence strengthens civilization

10.7.1 The Misconception of Modern Product Development

For decades, success was measured through speed:

- Faster releases
- Shorter sprint cycles
- Continuous deployment
- Rapid feature expansion
- Agile and DevOps enabled adaptability.

But in the AI era, speed alone cannot define success.

A faster product is not necessarily a better product.
A smarter system is not automatically a wiser system.

The challenge is not delivering intelligence quickly.
It is creating intelligence responsibly.

10.7.2 From Software Creation to Intelligence Creation

Traditional software executes predefined logic.

AI systems generate outcomes dynamically based on learned patterns.

Developers are no longer writing instructions alone.
They are shaping decision systems capable of influencing millions.

Every AI product implicitly answers:

- What decisions will this system influence?
- What behaviors will it encourage?
- What societal structures will it reinforce?
- What unintended consequences may emerge?

Product development becomes intelligence design.
Intelligence design carries responsibility.

10.7.3 Responsible Intelligence Creation

Responsible Intelligence Creation is central to the ZHA framework.

It aligns AI systems with:

- Human wellbeing
- Ecological balance
- Social harmony
- Long-term civilization stability

Responsible intelligence does not limit innovation.
It aligns innovation with purpose.

Example — Recommendation Algorithms

Traditional objective:
Increase engagement time.

Responsible objective:
Improve wellbeing and knowledge quality.

The same technology produces radically different outcomes depending on intention.

10.7.4 Innovation Must Remain Human-Centered

Innovation once focused on capability.
Human-centered innovation focuses on consequence.

Human-centered AI asks:

- Does this system empower people?

- Does it reduce cognitive overload?
- Does it enhance creativity?
- Does it preserve human agency?

Example — AI Coding Assistants

Technology-centric goal:
Automate development entirely.

Human-centered goal:
Support developers' creativity and problem-solving.

Outcome:
Capability expands without eroding human agency.

10.7.5 Technology Must Remain Sustainable

AI systems consume:

- Energy
- Infrastructure
- Computational resources
- Rare materials
- Responsible development includes sustainability awareness.

Example — Sustainable Model Engineering

Instead of endlessly scaling model size, organizations adopt:
Optimized architectures
Shared models
Efficient inference techniques
Innovation continues while environmental impact decreases.

10.7.6 Intelligence Must Strengthen Civilization

AI systems should strengthen:

- Education access
- Social equity
- Institutional trust
- Environmental resilience
- Human dignity

Example — AI Education Ecosystem

A civilization-aligned learning platform:

- Adapts to diverse learners
- Promotes critical thinking
- Encourages collaboration
- Supports underserved communities
- Technology becomes civilization infrastructure

10.7.7 The Evolution of Product Leadership

Historically optimized for:

- Timelines
- Feature adoption
- Revenue growth

AI-era product leaders must optimize for:

- Human impact
- Ethical alignment
- Sustainability outcomes
- Long-term societal stability

They evolve into AI Product Stewards.

Leadership shifts from management to stewardship.

10.7.8 AI Development as Continuous Responsibility

AI products never truly finish.

They evolve through:

- Model retraining
- Data updates
- Behavioral adaptation
- Societal interaction

Responsibility does not end at deployment.

Release becomes the beginning of stewardship.

10.7.9 The Competitive Advantage of Responsible Intelligence

Organizations practicing responsible intelligence gain:

- Regulatory resilience
- Customer trust
- Reduced ethical risk
- Sustainable innovation reputation
- Global leadership positioning

In the AI economy, trust becomes the most valuable asset.

Responsible intelligence builds trust at scale.

10.7.10 The ZHA Paradigm Shift

Old Paradigm:
Deliver software faster.

New Paradigm:

Create intelligence responsibly.

This transforms AI development from technological competition into collective human progress.

10.7.11 The Civilization Perspective

Humanity is entering the first era where intelligence itself is engineered.

AI creators are shaping:

- Social interaction
- Economic opportunity
- Cultural evolution
- Environmental sustainability

AI product development becomes one of the most consequential human activities ever undertaken.

Technology should not merely advance civilization.
It should help civilization endure.

Final Reflection

The future will not belong to those who build intelligence fastest.

It will belong to those who build intelligence wisely.

Responsible intelligence creation ensures that innovation strengthens humanity rather than overwhelms it.

CHAPTER XI

Chapter 11: AI Transformation Roadmap & Strategy

Choosing Sustainable SDLC Methodologies Powered by the ZHA Sustainable AI Framework

AI Transformation is often misunderstood as a technology program, selecting tools, building models, and automating operations. In the ZHA view, AI Transformation is a human-centric organizational evolution, designed to transition an enterprise from being digitally enabled to becoming intelligently guided while remaining aligned to sustainability, ethics, and long-term civilization value.

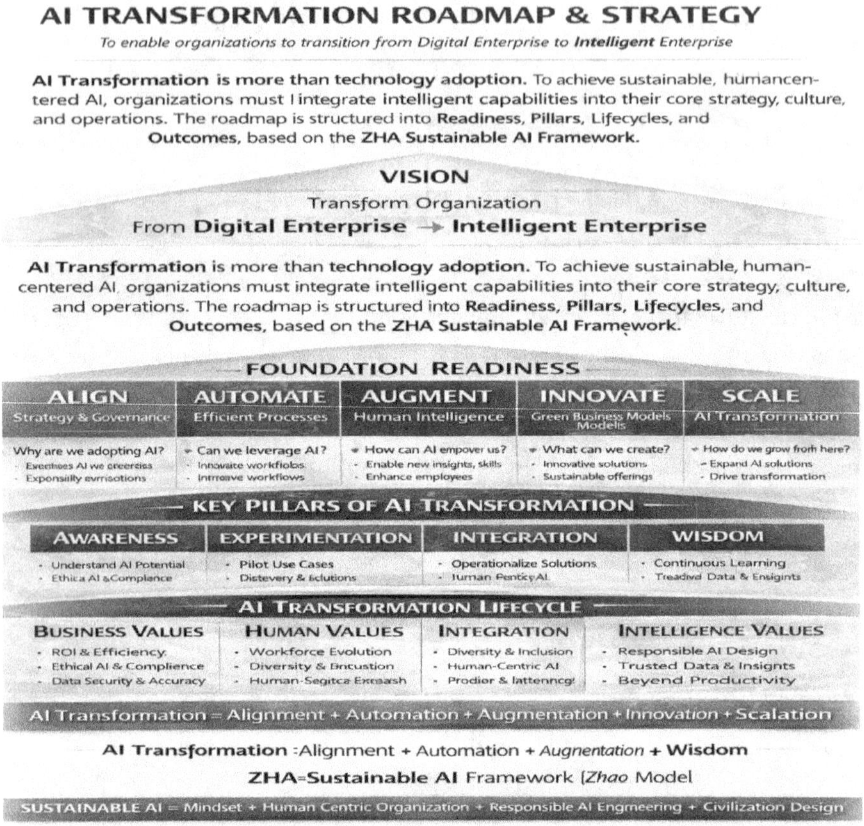

This chapter combines two inseparable decisions of the AI era:

- How an organization transforms (Roadmap & Strategy)
- How an organization builds intelligence (Sustainable SDLC Selection)

In the AI era, SDLC is no longer merely a delivery method—it becomes the operating philosophy for creating intelligence responsibly.

11.1 The SDLC Decision Crisis in the AI Era

Traditionally, Software Development Life Cycle (SDLC) methodologies were selected based on operational needs:

- Speed of delivery
- Project complexity
- Team size
- Regulatory constraints

Organizations debate:

- Agile vs. Waterfall
- DevOps vs. Traditional delivery
- Scrum vs. Kanban
- SAFe vs. Hybrid

However, Artificial Intelligence fundamentally changes the decision.

AI systems do not merely deliver software. They influence:

- Human decisions
- Social structures
- Economic opportunities
- Environmental sustainability

Selecting an SDLC methodology therefore becomes a civilizational decision, not merely an engineering choice.

The central ZHA insight:

SDLC must align with Sustainable AI outcomes, not only delivery efficiency.

11.2 AI Transformation Vision: From Digital Enterprise to Intelligent Enterprise

Transform Organization → From Digital Enterprise → Intelligent Enterprise

AI Transformation is not technology adoption.

It is organizational evolution combining:

- Strategy
- Culture
- Data
- Automation
- Human Intelligence + Machine Intelligence
- ZHA Elevation

AI Transformation must be guided by a Sustainability Mindset to ensure long-term civilization value—not merely short-term enterprise gains.

11.3 The ZHA Sustainable AI Architecture Within AI Transformation

The AI Transformation Roadmap operates inside the ZHA Sustainable AI Framework:

- Layer 1: Sustainability Mindset Foundation
- Layer 2: Human-Centric AI Organization
- Layer 3: Sustainable AI Engineering
- Layer 4: AI Civilization Transformation

Outcome: Civilization Values

AI Transformation is not just a maturity journey of technology—
it is a maturity journey of:

Mindset → Organization → Engineering → Civilization Stewardship

11.4 The AI Transformation Pyramid

AI Transformation can be understood as a pyramid of readiness, execution, and outcomes:

OUTCOMES (Value Creation)
→ Intelligent Enterprise

AI Transformation Pillars
→ Data | People | Process | Technology

Foundations (Readiness Layer)
→ AI readiness determines success or failure

Alignment with ZHA:

- Mindset enables readiness
- Organization design enables governance and adoption
- Engineering enables scalable AI systems
- Civilization layer ensures long-term stewardship

11.5 Foundation Layer: AI Readiness (Where Transformation Succeeds or Fails)

Before transformation begins, organizations must become AI-Ready.

11.5.1 Leadership & Vision Alignment

- AI vision linked to business strategy
- Executive sponsorship
- Value-driven transformation
- Responsible AI governance

Guiding Question: Why are we adopting AI?

ZHA Link: Conscious Awareness + Responsible Creation
AI must serve human dignity, planetary balance, and shared prosperity.

11.5.2 Data Foundation

- Data governance
- Data quality & ownership
- Unified data architecture
- Security & privacy
- Metadata & knowledge models

Principle: AI runs on data maturity.
ZHA Link: Accountability of Intelligence

Sustainable AI requires ethical, transparent, purpose-driven data ecosystems.

11.5.3 Culture & Mindset

- AI literacy across workforce
- Learning organization mindset
- Experimentation culture
- Human–AI collaboration acceptance

Principle: AI Transformation = Mindset Transformation

Without cultural evolution, automation creates resistance.
With mindset evolution, AI creates augmentation.

11.5.4 Digital & Technology Readiness

- Cloud infrastructure
- APIs & integration layer
- Automation platforms
- Cybersecurity maturity

Technology must be scalable and secure—guided by sustainability principles.

11.5.5 Ethical & Responsible AI

- Bias control
- Transparency
- Compliance framework
- Trust architecture

ZHA Principle: Responsible AI protects. Sustainable AI guides.

Foundation Outcome: AI-Ready Organization

11.6 Key AI Transformation Pillars (Execution Model)

These pillars operationalize AI adoption inside the ZHA framework.

11.6.1 ALIGN — Strategy & Governance
Purpose: Connect AI initiatives to business outcomes
Capabilities:

- Roadmap
- Prioritization
- ROI governance
- Portfolio management
- Responsible AI policies

Outcome: AI aligned to enterprise strategy and sustainability values

ZHA Link: Human-Centric AI Organization

11.6.2 AUTOMATE — Intelligent Operations

Purpose:

- Remove repetitive work

Capabilities:

- Process mining

- RPA + AI
- Intelligent workflows
- Decision automation
- AI agents

Outcome: Efficiency + cost optimization + reduced burnout

ZHA Link: Sustainable AI Engineering

Automation must respect:

- Human dignity
- Workforce transition
- Ethical deployment

11.6.3 AUGMENT — Human Intelligence Amplification

Purpose: Enhance human capability, not replace it.

Capabilities:

- GenAI copilots
- Knowledge assistants
- Decision support
- Predictive insights

Outcome: AI-augmented professionals

ZHA Link: Human–AI Collaboration

11.6.4 INNOVATE — AI-Driven Business Models

Purpose: Create new sustainable value streams

Capabilities:

- AI products/services
- ESG analytics
- Data monetization
- Ecosystems
- Autonomous systems

Outcome: New revenue aligned with long-term prosperity

ZHA Link: Regenerative Stewardship

11.6.5 SCALE & SUSTAIN — Enterprise Adoption

Purpose: Move from pilots to enterprise impact

Capabilities:

- AI operating model
- MLOps / LLMOps
- AI CoE
- Reusable assets
- Continuous learning loops

Outcome: Sustainable AI transformation at scale

ZHA Link: AI Civilization Transformation Layer

11.7 Why SDLC Becomes the Core Engine of AI Transformation

Traditional Environment:
SDLC = Project Execution Framework

AI Era:
SDLC = Organizational Transformation Framework

The SDLC you choose determines:

- How intelligence is created
- How risks are governed
- How humans interact with automation
- How sustainability is preserved

SDLC becomes a reflection of values—not merely delivery.

11.8 Sustainable SDLC Principles (ZHA Perspective)

Before selecting any methodology, establish five sustainability principles:

- Human-Centricity — Technology must augment people
- Ethical Governance — AI decisions remain accountable
- Environmental Responsibility — Infrastructure minimizes ecological impact
- Social Inclusion — Innovation benefits broad society
- Long-Term Resilience — Systems remain maintainable for decades

These form the foundation of Sustainable SDLC.

11.9 Mapping SDLC Methodologies to AI Transformation Needs

No single SDLC fits all organizations. ZHA introduces contextual selection based on maturity and responsibility.

11.9.1 Agile SDLC — Human-Centric Innovation

Strengths:

- Iterative learning
- rapid feedback
- adaptability

Sustainable AI Application: Continuous human-impact evaluation

Example:

- AI Learning Platform
- Test personalization models
- Incorporate teacher feedback
- Refine ethical data usage policies

Include sustainability acceptance criteria per sprint

ESG Alignment:

- Social inclusion
- ethical improvement
- user wellbeing

11.9.2 DevOps & MLOps — Responsible Automation

DevOps enables continuous delivery; AI introduces MLOps for model lifecycle.

Sustainable DevOps Principles:

- Energy-efficient pipelines
- Bias monitoring automation
- Responsible rollback mechanisms
- Explainability monitoring
- Example: Financial Fraud Detection
- Fairness testing
- Regulatory validation
- Environmental compute monitoring

ESG Alignment: Governance, risk management, operational sustainability

11.9.3 Waterfall SDLC — High-Regulation Environments

Sequential governance where predictability and accountability are critical.

Example:

- Government Healthcare AI
- Strict validation
- policy approval
- citizen data protection

ESG Alignment: Transparency, accountability, public trust

11.9.4 SAFe — Enterprise AI Transformation

Coordinates multiple teams with business alignment.

Example:

- Global Manufacturing Enterprise
- Supply chain AI
- robotics
- sustainability monitoring

ESG Alignment: Enterprise sustainability strategy, resilience

11.9.5 Hybrid Sustainable SDLC (ZHA Recommended)

AI transformation often requires:

- Agile for innovation
- DevOps/MLOps for lifecycle automation
- Waterfall gates for governance
- SAFe for enterprise coordination

This becomes Sustainable SDLC Architecture aligned to transformation stage.

11.10 Embedding ESG Into the SDLC (From Intent to Execution)

Traditional metrics: Scope, schedule, cost.
AI-era metrics: Scope + Schedule + Cost + ESG impact.

A Sustainable SDLC integrates ESG as non-negotiable quality gates.

Practical Implementation:

- ESG Definition of Done (DoD)
- ESG stage-gates (Plan → Build → Test → Release → Operate)
- ESG evidence artifacts stored with versioning

Environmental (E): Carbon-Intelligent Engineering

Focus: Compute, energy, infrastructure optimization.

SDLC Insertions:

- Carbon-aware architecture decisions
- Energy-efficient training practices
- Compute intensity measurement
- Right-sized deployment
- Continuous energy monitoring
- Environmental Checkpoints
- Carbon-aware infrastructure
- Compute efficiency targets
- Sustainable cloud strategy
- Model lifecycle controls

Social (S): Inclusive, Human-Centric AI Delivery

Focus: Fair distribution, accessibility, workforce transition.

SDLC Insertions:

- Harm/benefit analysis in requirements
- Dataset diversity checks
- Fairness + usability testing
- Human escalation workflows
- Continuous feedback loops

Governance (G): Accountability Built Into Release

Governance is integrated—not external.

SDLC Insertions:

- Risk classification
- Decision traceability
- Explainability testing
- Governance sign-offs
- Drift monitoring
- Governance Gates
- Bias thresholds met
- Privacy requirements met
- Documentation attached
- Risk owner approval
- Rollback plan verified
- Practical ESG Gates Across the SDLC
- StageESG Gate

PlanESG impact statement + risk classification

BuildEfficiency targets + inclusion checks

TestFairness + accessibility + explainability evidence

ReleaseGovernance approval + audit trail + rollback plan

OperateDrift monitoring + carbon + human impact + compliance

Result: ESG becomes a measurable engineering discipline—not a late-stage audit.

SDLC Phase	Environmental (E)	Social (S)	Governance (G)	Evidence Artifacts	Owner Role
1. Plan / Discover	Carbon-aware architecture selection Compute intensity estimation Cloud region impact analysis	Stakeholder impact mapping Inclusion risk assessment Workforce impact estimation	AI risk classification Regulatory applicability check Accountability owner assigned	ESG Impact Statement Risk Register Entry Architecture Decision Record Sustainability Charter	Product Owner AI Architect Chief Sustainability Officer AI Governance Lead
2. Design	Model size optimization strategy Energy-efficient training design Sustainable infrastructure planning	Dataset diversity planning Accessibility requirements defined Human-in-the-loop design decisions	Data lineage mapping Privacy & compliance validation Explainability approach defined	Data Governance Document Model Design Specification Inclusion & Accessibility Plan Governance Checklist	Data Lead ML Architect UX Lead Compliance Officer
3. Build / Develop	Efficient training methods Compute usage logging Resource optimization controls	Bias mitigation techniques Inclusive dataset validation Workforce transition communication	Model documentation Version control traceability Secure coding standards	Model Card Dataset Audit Report Compute/Energy Log Code Repository Audit Trail	ML Engineer Data Scientist DevOps Lead Security Lead
4. Test / Validate	Energy consumption benchmarking Performance vs carbon trade-off analysis	Fairness testing Accessibility validation Usability testing	Explainability review Ethical risk reassessment Regulatory validation	Fairness Test Report Explainability Report Ethical Risk Assessment Compliance Sign-off	QA Lead Responsible AI Officer Governance Committee
5. Release / Deploy	Carbon-efficient deployment region Right-sized inference infrastructure	Transparency notices Escalation channels enabled	Governance gate approval Rollback strategy validated	Release Approval Record Deployment Audit Log Risk Acceptance Document Rollback Plan	Release Manager AI Governance Board CTO
6.	Ongoing	Bias drift	Continuous	Drift	MLOps Lead

Operate / Monitor	carbon monitoring Compute optimization adjustments	monitoring Workforce reskilling tracking Human oversight logs	compliance monitoring Incident response workflow Periodic governance review	Monitoring Dashboard ESG Performance Report Incident Postmortem Continuous Audit Log	Sustainability Team AI Ethics Committee Board Oversight

11.11 SDLC Selection Based on AI Maturity Levels

ZHA positions mature organizations at the Civilization Stewardship level.

11.12 AI Transformation Lifecycle With SDLC Alignment

As organizations evolve through transformation stages, SDLC evolves too:

- Awareness → Agile pilots + literacy + governance basics
- Experimentation → Agile + responsible data + measurement
- Operationalization → DevOps/MLOps + release governance
- Intelligence → scalable SDLC architecture + continuous monitoring
- Wisdom / Autonomous → ZHA Sustainable SDLC + civilization values enforcement

This is how SDLC becomes the mechanism that turns roadmap intent into real transformation.

11.13 Organizational Transformation Practices Required for Sustainable SDLC

Choosing Sustainable SDLC requires organizational evolution:

- Leadership Transformation: delivery focus → stewardship focus

- Cultural Transformation: experimentation with responsibility
- Skill Transformation: ethics, sustainability, systems thinking
- Governance Transformation: decisions include societal impact

11.14 Real-Time Enterprise Transformation Scenario

Phase 1 – Alignment
Define sustainability goals and AI vision.

Phase 2 – Methodology Selection
Choose hybrid SDLC aligned with maturity.

Phase 3 – Implementation
Embed ESG metrics into backlog and pipelines.

Phase 4 – Continuous Governance
Monitor societal impact continuously.

Result: AI transformation becomes stable, ethical, and sustainable.

11.15 Value Creation Outcomes (ZHA Multi-Dimensional Impact)

- AI Transformation under ZHA creates value across four dimensions:
- Business: faster decisions, cost optimization, innovation acceleration
- Human: creativity, skill evolution, reduced burnout
- Organizational: agility, resilience, transparency, learning culture
- Societal: responsible technology, sustainable growth, inclusive innovation

ZHA differentiation: transformation is not only enterprise-centric—it is civilization-centric.

11.16 AI Transformation Value Equation (ZHA Extended)

AI Value = (Aligned Strategy) + (Trusted Data) + (Empowered People) + (Intelligent Automation) + (Scalable Governance)

ZHA Extended View:
AI Value = Traditional value equation + Sustainability Mindset

Sustainable AI = Mindset + Human-Centric Organization + Responsible AI Engineering + Civilization Design

11.17 Why Sustainable SDLC Matters for the Future

The next generation of global regulations will evaluate organizations not only by profitability but by responsibility. Sustainable SDLC prepares organizations for:

- responsible AI regulations
- ESG investment expectations
- societal trust requirements
- long-term innovation resilience

Organizations adopting Sustainable SDLC early become global leaders.

11.18 The ZHA Insight

The most successful organizations of the AI era will not be those that deliver software fastest.
They will be those that deliver intelligence responsibly, sustainably, and human-centrically.
Choosing an SDLC methodology therefore becomes an ethical decision shaping civilization itself.

Chapter Reflection

Methodologies do not transform organizations.
Mindsets guiding methodologies do.

When AI Transformation Roadmap & Strategy is executed through Sustainable SDLC, Artificial Intelligence evolves as a force for long-term human progress—aligned to ESG, sustainability values, and civilization outcomes.

CHAPTER XII

Chapter 12: Protecting ESG During Technological Disruption

Describes organizational transformation practices aligning Artificial Intelligence with ESG, sustainability values, and long-term societal impact.

12.1 The Paradox of Technological Progress
Technological disruption has always promised progress.

Artificial Intelligence accelerates innovation across industries:

- Automated healthcare diagnostics
- Intelligent financial systems
- Smart cities
- Autonomous logistics
- Personalized education

Yet history reveals a recurring pattern:

Every major technological leap creates unintended consequences when adoption outpaces responsibility.
Industrialization increased productivity but damaged ecosystems.
Digital platforms connected humanity but amplified misinformation.
Automation improved efficiency while destabilizing employment structures.
AI introduces disruption at an unprecedented scale.

The Central Question:
How can organizations innovate rapidly while protecting environmental, social, and governance stability?

Answer: Integrate ESG protection directly into AI transformation.

12.2 Understanding ESG in the AI Era

ESG — Environmental, Social, and Governance — is often misunderstood as reporting compliance.

In reality, ESG represents civilization stability metrics.

Environmental (E)
How technology impacts planetary health.

Social (S)
How innovation affects people, communities, and equality.

Governance (G)
How decisions remain ethical, transparent, and accountable.

Unlike traditional technologies, AI influences ESG outcomes continuously after deployment. Therefore, ESG must move from reporting practice to design principle.

12.3 Technological Disruption as an ESG Risk Multiplier

Artificial Intelligence magnifies both positive and negative impact.

Without ESG integration, disruption creates systemic risks:

- Environmental Risks
- High energy consumption
- Electronic waste from hardware cycles
- Social Risks
- Workforce displacement
- Algorithmic bias
- Digital inequality
- Governance Risks
- Opaque decision-making
- Loss of accountability
- Concentration of technological power

AI becomes a risk multiplier unless guided by sustainability mindset principles.

12.4 Organizational Transformation for ESG Protection

Protecting ESG during disruption requires transformation across four dimensions:

- Leadership mindset
- Operational practices
- Technology governance
- Cultural evolution

12.5 Leadership Transformation: From Profit to Stewardship

Traditional leadership prioritizes financial performance.
AI-era leadership must adopt stewardship responsibility.

Leaders evaluate decisions through three lenses:

- Business success
- Societal impact
- Planetary sustainability

Example — Retail AI Transformation

Profit-Focused Approach:

- Maximize sales volume.
- Stewardship Approach:
- Reduce overproduction
- Minimize waste
- Optimize logistics emissions

Outcome: Higher efficiency with reduced environmental impact.

Leadership intent determines ESG outcome.

12.6 Environmental Protection During AI Transformation

AI infrastructure consumes significant energy resources.

Organizations must design environmentally responsible intelligence systems.

Sustainable Practices

- Energy-efficient model design
- Green cloud infrastructure
- Carbon-aware computing
- Responsible hardware lifecycle management

Example — Sustainable Data Centers

AI training workloads scheduled when renewable energy availability is highest.

Results:

- Reduced carbon emissions
- Optimized energy usage
- Regulatory readiness

AI becomes part of the climate solution rather than climate burden.

12.7 Social Protection During Technological Disruption

AI transformation often creates fear around job displacement and inequality.

Technology must elevate society—not destabilize it.

Workforce Transformation Example: Manufacturing introduces robotics.

Instead of layoffs:

- Reskilling programs
- Digital education initiatives
- AI supervision roles
- Workers evolve alongside automation.

Digital Inclusion Example: AI education platforms expand access to underserved communities.

Technology reduces inequality rather than widening gaps.

12.8 Governance Protection in AI Systems

Governance is the most critical ESG pillar in the AI era.

AI decisions must remain explainable and accountable.

- Governance Practices
- Responsible AI review boards
- Algorithm transparency policies
- Human oversight mechanisms
- Ethical risk assessments
- Continuous auditing

Example — Financial AI Governance

AI credit scoring includes:

- Bias monitoring dashboards
- Regulatory compliance checkpoints
- Human override controls

Trust increases among customers and regulators. Governance becomes an innovation enabler.

12.9 Embedding ESG into the AI Product Lifecycle

ESG protection must exist across every stage:

Alignment Stage

Define sustainability objectives.

Design Stage

Architect ethical safeguards (e.g., privacy-by-design structures).

Development Stage

Monitor energy usage and dataset fairness.

Deployment Stage

Introduce human oversight and community consultation.

Operations Stage

Continuous ESG monitoring dashboards. ESG becomes operational—not ceremonial.

12.10 ESG Metrics for AI Transformation

Measuring Responsible Intelligence

Organizations must measure ESG continuously.

12.10.1 Why ESG Measurement Is Critical

AI systems:

- Learn from data
- Adapt behaviors
- Influence decisions
- Reshape economic and social patterns
- Impact is ongoing.

Without measurement:

- Environmental harm grows unnoticed
- Social harm accumulates silently
- Governance risks emerge too late

ZHA Principle: What gets measured becomes sustainable.

12.10.2 Environmental Metrics

1. Carbon Footprint per Model
   Tracks emissions from training, fine-tuning, and inference.
   Each release includes:

- Carbon impact report
- Sustainability score

2. Energy Consumption
Tracks electricity usage across:

- GPUs
- Data centers
- Storage systems

Workloads can shift to renewable-powered regions.

3. Infrastructure Efficiency
Measures:

- Compute utilization
- Storage optimization
- Cooling efficiency

Environmental transformation outcome:
AI becomes climate-aware technology.

12.10.3 Social Metrics

1. Workforce Impact

Measures:

- Roles automated
- Employees reskilled
- New roles created
- Satisfaction levels

2. Inclusion Indicators

Evaluates:

- Bias detection
- Diversity representation
- Equitable outcomes

3. Accessibility Adoption

Ensures usability for:

- Differently-abled users
- Low digital literacy communities
- Rural populations

Social transformation outcome:
AI expands opportunity instead of deepening inequality.

12.10.4 Governance Metrics

1. Algorithm Transparency Score
Measures explainability and decision traceability.

2. Audit Compliance Rate
Tracks adherence to ethical and regulatory standards.

3. Ethical Incident Tracking
Monitors bias incidents, privacy violations, harmful outcomes.

Governance transformation outcome:
Shift from compliance culture to responsibility culture.

12.10.5 Integrated ESG Dashboard

Leading organizations build real-time ESG dashboards reviewed alongside financial performance.

Sustainability becomes strategic.

12.10.6 ESG Across the AI Lifecycle

- Alignment → Define ESG targets
- Design → Architect sustainable solutions
- Development → Track metrics
- Deployment → Monitor societal impact
- Operations → Continuous improvement

ESG becomes embedded into SDLC.

12.10.7 Competitive Advantage of ESG Measurement

Effective ESG measurement drives:

- Investor confidence
- Regulatory readiness

- Talent attraction
- Customer loyalty
- Long-term innovation stability

Responsible AI becomes strategic advantage.

12.10.8 The ZHA Insight

AI amplifies human capability.
ESG ensures it amplifies human responsibility.

Measurement transforms sustainability from intention into practice.

12.11 The Role of Culture in AI ESG Protection

Why Sustainable AI Depends on Organizational Mindset

Policies alone cannot protect sustainability during rapid disruption.

AI evolves quickly. Culture becomes the true decision-maker.

Employees make daily decisions no policy can supervise.

ESG protection ultimately depends on shared values.

Core Cultural Pillars

1. Responsibility Culture
Ownership of technological impact.

2. Transparency Culture
Openness about risks, uncertainty, limitations.

3. Collaboration Culture
Cross-functional and societal partnership.

4. Long-Term Thinking Culture

Decisions evaluated across decades—not quarters.

12.11.9 Culture as the Ultimate AI Safety Mechanism

Technical safeguards may fail.

Regulations may lag.

A strong responsibility culture continuously corrects course.

Culture becomes the most resilient safeguard.

12.12 ESG as Competitive Advantage

ESG strengthens long-term competitiveness through:

- Investor confidence
- Regulatory resilience
- Customer trust
- Talent attraction
- Brand legitimacy

Responsible organizations outperform disruptive ones over time.

12.13 The ZHA Framework for AI ESG Protection

From Sustainability Reporting to Responsible Intelligence Governance

Traditional ESG is reactive.

The ZHA model makes ESG proactive and mindset-driven.

Under ZHA:

- ESG is not reporting.
- ESG becomes decision consciousness.

ZHA AI ESG Evolution Model

Mindset → Organization → Civilization
Compliance → Responsibility → Stewardship

12.14 From Disruption to Sustainable Civilization

Transforming Technological Shock into Human Progress

AI represents the most powerful disruption in history because it amplifies intelligence itself.

Humanity faces two paths:
Path One — Accelerated Instability

Unchecked automation, inequality, environmental strain.

Path Two — Sustainable Global Progress
AI aligned with ESG strengthens civilization.

AI ESG functions as civilization stabilization architecture.

Without ESG, intelligence scales risk.
With ESG, intelligence scales responsibility.

Chapter Reflection

SDLC must align with Sustainable AI outcomes—not only delivery efficiency.
Innovation without ESG protection creates temporary success.
Innovation aligned with sustainability creates enduring civilization.

The future will belong not to the fastest innovators,
but to the wisest stewards —
organizations that protect people, planet, and governance
while advancing intelligence for humanity.

CHAPTER XIII

Chapter 13 - Responsible Innovation for Sustainable AI Transformation Governance

Aligning Students, Professionals, and Business Leaders for a Human-Centric Intelligent Civilization

13.1 The Innovation Crisis of the AI Era

Humanity has never lacked innovation.

What humanity increasingly lacks is responsible innovation.

Artificial Intelligence enables individuals and organizations to create powerful systems rapidly:

- autonomous decision engines,
- generative intelligence platforms,
- predictive governance systems,
- algorithmic economic models.

Yet innovation often advances faster than wisdom.

Technological breakthroughs appear before society fully understands their consequences.

This creates an innovation gap:
Capability grows faster than responsibility.

Responsible Innovation closes this gap.

It ensures that innovation strengthens civilization rather than destabilizes it.

13.2 What Is Responsible Innovation?

Responsible Innovation is the intentional alignment of innovation with:

- human wellbeing,
- environmental sustainability,
- ethical governance,
- long-term societal benefit.

It asks innovators to consider not only:

- Can we build this?

but also:

- Should we build this, and how should it serve humanity?

Responsible innovation integrates foresight, ethics, and sustainability into every stage of creation.

13.3 Sustainable AI Governance - A New Global Requirement

From Compliance Governance to Civilization Governance

Traditional governance models were designed for predictable technologies.

Organizations historically governed innovation through:

- regulations,
- internal policies,
- risk mitigation frameworks,
- compliance audits.

These mechanisms assume technology behaves deterministically.

Artificial Intelligence changes this assumption.

AI systems:

- learn continuously,
- evolve after deployment,
- influence human decisions,
- reshape markets and societies dynamically.

Therefore, governance based only on compliance becomes insufficient.

The Limitation of Traditional AI Governance

Compliance governance asks:

- Are we legally safe?
- Did we meet regulatory requirements?
- Are risks documented?

But Sustainable AI Governance asks deeper questions:

- Is this technology socially beneficial?
- Does it preserve human dignity?
- Will it remain ethical over decades?
- Does it strengthen civilization stability?

Sustainable AI Governance Defined

Sustainable AI Governance creates innovation ecosystems where technology remains:

- Transparent
- Users understand how AI decisions are made.

Example - AI Loan Approval

Instead of opaque scoring models, organizations provide explainable reasoning:

- credit factors shown,
- appeal mechanisms enabled,
- fairness audits conducted.
- Transparency builds trust.
- Accountable

Organizations remain responsible for AI outcomes.

Example - Autonomous Vehicles
Responsibility frameworks define:

- developer accountability,
- operational oversight,
- incident response governance.
- AI cannot exist without human accountability.
- Inclusive

Innovation benefits diverse populations.

Example - Healthcare AI
Datasets include multiple demographic groups to avoid biased medical predictions.

Inclusivity prevents technological inequality.

Beneficial Across Generations

Decisions consider long-term societal impact.

Example - Data Infrastructure
Instead of maximizing short-term data monetization, organizations adopt:

- privacy-by-design,

- sustainable data retention,
- ethical data usage models.

Technology serves future generations.

Governance as Trust Infrastructure

Under the ZHA model:

- Governance is not a restriction on innovation.
- It becomes the foundation enabling innovation adoption.

When citizens trust technology:

- adoption accelerates,
- resistance decreases,
- innovation becomes sustainable.

13.4 The ZHA Responsible Innovation Ecosystem

Innovation as a Collective Responsibility

Modern innovation no longer belongs to isolated organizations.

AI influences entire civilizations.

ZHA therefore introduces a multi-stakeholder innovation ecosystem aligning four essential actors.

1. Students - Future Creators

Students shape future technological culture.
They represent pre-industry governance formation.

2. Professionals - System Builders

Professionals operationalize innovation daily.
Their decisions determine whether AI becomes ethical or harmful.

3. Business Leaders - Direction Setters

Leadership defines incentives and strategic priorities.
They determine whether innovation pursues exploitation or sustainability.

4. Society & Institutions - Civilization Stabilizers
Governments, universities, and communities ensure innovation aligns with public interest.

ZHA Ecosystem Insight

Responsible innovation succeeds only when all four groups evolve simultaneously.

If one group advances without others:

- technology outpaces society,
- governance becomes reactive,
- instability increases.

Real Example - Smart City Deployment

Successful smart cities involve:

- students researching sustainability,
- professionals designing systems,
- leaders funding ethical innovation,
- citizens participating in governance.

Innovation becomes shared ownership.

13.5 Students as Future Responsible Innovators

Education as Civilization Infrastructure

Students represent the origin point of long-term transformation.

Traditional education emphasizes:

- technical knowledge,
- academic achievement,
- examination performance.

Responsible innovation education develops mindset maturity.

Four Capabilities Students Must Learn

1. Sustainability Thinking
Understanding environmental and social consequences.

Example
Students analyze carbon impact of AI models during coursework.

2. Ethical Technology Design
Embedding fairness into creation.

Example
Student AI projects include bias evaluation and ethical review sections.

3. Collaborative Problem Solving
Innovation becomes interdisciplinary.

Example
Engineering students collaborate with social science and environmental departments.

4. Societal Responsibility
Students see technology as public service.

Example - ZSPC Student Model

Through ZSPC certification programs students:

- Conduct environmental sustainability projects.
- Design local community solutions.
- Explore AI's societal implications.
- Learn governance awareness early.

Outcome

Students become:

- innovation citizens
- ethical creators
- sustainability practitioners

—not passive technology consumers.

AI Governance Impact

When such students enter industry:

- ethics becomes instinctive,
- governance becomes culture,
- regulation becomes secondary reinforcement.

This is preventive governance.

13.6 Professionals as Responsible Practitioners

The Execution Layer of Civilization

Professionals translate ideas into deployed systems.

Every algorithm, product feature, and deployment decision passes through professionals.

Therefore professionals are civilization gatekeepers.

Evolution of Professional Identity

Traditional Professional

Focus:

- efficiency,
- performance,
- delivery speed.

Responsible Practitioner

Focus expands to include:

- Ethical judgment
- Sustainability awareness
- Human-centered decision-making

Example - Software Engineer Transformation

Traditional Engineer:

- optimizes model accuracy,
- improves processing speed.

Responsible Innovator, also evaluates:

- dataset bias,
- environmental compute cost,
- societal influence of outputs.

- Engineering becomes stewardship.
- Workplace Impact

Organizations adopting responsible innovation practices experience:

- Improved collaboration → Teams align around shared purpose.
- Reduced ethical risk → Fewer reputational crises.
- Stronger innovation culture → Employees innovate responsibly.

Example - AI Development Team

Before release, teams conduct:

- ethical impact reviews,
- ESG evaluation,
- stakeholder feedback sessions.

Innovation quality improves alongside trust.

13.7 Business Leaders as Civilization Stewards

Leadership in the Age of Artificial Intelligence

Business leaders now hold unprecedented power.

A single executive decision can influence millions of lives through AI deployment.

Historically leadership optimized:

- profit,
- growth,
- market dominance.

The AI era requires a shift toward Stewardship Leadership.

Responsible Innovation Leadership Model

Leaders must balance three forces:

Example - AI Product Strategy Decision

A technology company considers predictive behavioral advertising.

Short-Term Profit Model:

- maximize engagement,
- exploit psychological patterns,
- increase addiction-based usage.

Result:
Revenue growth but declining societal trust.

Responsible Innovation Model

Leadership introduces:

- algorithm transparency,
- user control options,
- limits on manipulative personalization,
- wellbeing-focused metrics.

Outcome:
Lower short-term gain, higher long-term trust and brand resilience.

Trust becomes competitive advantage.

Leadership Transformation

Leadership evolves from:
Owner → Operator → Steward

Stewards recognize that:

Technology decisions shape civilization outcomes.
Organizations become custodians of societal wellbeing.

ZHA Leadership Insight

In the Sustainable AI era:
The most successful leaders will not be those who build the fastest technologies,
but those who guide technology toward human flourishing.

13.8 Universities and Institutions as Governance Foundations

Educational institutions and research organizations play a critical role in sustainable AI governance.

They shape:

- ethical research practices,
- interdisciplinary collaboration,
- future policy thinking.

Example - Responsible AI Curriculum

Universities integrate courses combining:

- AI engineering,
- environmental science,
- ethics,
- social policy.

Graduates enter industry already aligned with sustainability governance.

13.9 Businesses and Industry as Responsible Innovation Engines

Organizations must move beyond innovation labs focused solely on technological novelty.

Responsible innovation engines integrate:

- ESG strategy,
- sustainability KPIs,
- ethical review frameworks,
- stakeholder participation.

Example - Smart Manufacturing

AI-driven manufacturing optimizes production while:

- reducing emissions,
- improving worker safety,
- conserving resources.

Innovation produces both economic and environmental value.

13.10 Sustainable AI Governance Framework

ZHA proposes governance operating across three layers:

Layer 1 - Individual Responsibility
Ethical awareness among creators.

Layer 2 - Organizational Governance
Policies, audits, and accountability mechanisms.

Layer 3 - Societal Alignment
Collaboration between governments, academia, and communities.

Governance becomes distributed rather than centralized.

13.11 Responsible Innovation Lifecycle

Responsible innovation follows a continuous cycle:

- Anticipate Impact
- Reflect on Ethics
- Engage Stakeholders
- Act Responsibly
- Learn and Adapt

?

Innovation becomes a learning system.

Example - AI Healthcare Platform

Before deployment:

- patients consulted,
- clinicians involved,
- ethical review conducted.

After deployment:

- outcomes monitored,
- bias corrected,
- system continuously improved.

Governance evolves alongside innovation.

13.12 Aligning Innovation with UN Sustainable Development Goals

Transforming Artificial Intelligence into a Force for Global Progress

13.12.1 Innovation and the Global Responsibility Era

The United Nations Sustainable Development Goals (SDGs) represent humanity's collective agreement on the future we want to build.

The SDGs address humanity's most urgent challenges:

- education inequality,
- climate crisis,
- economic imbalance,
- unsustainable consumption,
- rapid urbanization.

Artificial Intelligence now sits at the center of these challenges.

AI can either:

accelerate inequality and environmental damage, or become humanity's most powerful tool for sustainable development.

Responsible innovation ensures AI advances global wellbeing, not only technological capability.

Under the ZHA Sustainability Mindset, innovation automatically aligns with SDGs because innovation decisions consider:

- people,
- planet,
- prosperity,
- partnerships.

AI shifts from disruption engine to global progress infrastructure.

13.12.2 SDG 4 - Quality Education

AI as a Universal Learning Equalizer

Quality education is the foundation of sustainable civilization.

Traditional education systems face limitations:

- unequal access,
- resource shortages,
- geographic barriers,
- outdated curriculum models.

AI enables personalized and scalable learning experiences.

Responsible AI Application
Example - AI Learning Platforms

AI tutors adapt learning to each student's pace:

- rural students receive advanced education access,
- language barriers reduced through AI translation,
- adaptive learning supports differently-abled students.

ZHA Alignment

Programs like ZSPC demonstrate how education evolves beyond knowledge transfer into mindset development:

Students learn:

- sustainability thinking,
- ethical technology awareness,
- civic responsibility.

Outcome:
Education produces responsible innovators rather than only skilled workers.
AI strengthens human potential.

13.12.3 SDG 13 - Climate Action

AI as Environmental Intelligence

Climate change represents the greatest collective risk facing humanity.

AI can dramatically improve environmental protection.

Responsible AI Application

Example - Climate Prediction Systems

AI analyzes satellite data to:

- predict floods,
- forecast drought patterns,
- optimize renewable energy distribution.

Governments act proactively instead of reactively.

Example - Sustainable Computing

Organizations following ZHA principles adopt:

- energy-efficient AI models,
- optimized data centers,
- carbon-aware computing schedules.

Innovation reduces environmental impact.

ZHA Insight

Feeling Sufficient + Resource Conservation ensures AI growth does not become ecological exploitation.

Technology becomes planetary stewardship.

13.12.4 SDG 10 - Reduced Inequality

Inclusive Intelligence for All

Technological revolutions historically increase inequality before stabilizing.

AI risks concentrating power among:

- wealthy nations,
- large corporations,
- technologically advanced regions.

Responsible innovation distributes opportunity.

Responsible AI Application

Example - AI Agriculture Advisory

Small farmers access AI-driven crop insights through mobile platforms:

- improved yields,
- reduced resource waste,
- economic empowerment.

Technology reaches underserved communities.

Example - Accessible Healthcare AI
AI diagnostics bring specialist-level healthcare to remote regions lacking medical experts.

AI becomes equalizer rather than divider.

ZHA Alignment

Principles supporting equality:

- Social Equality
- Gender Equality
- Community Engagement
- Sharing & Caring Collaboration

Innovation expands opportunity across society.

13.12.5 SDG 12 - Responsible Consumption and Production

Intelligent Sustainability in Economic Systems

Modern economies encourage overproduction and overconsumption.

AI can either intensify consumption patterns or optimize resource use.

Responsible innovation transforms production systems.

Responsible AI Application

Example - Smart Supply Chains

AI predicts demand accurately, enabling:

- reduced manufacturing waste,
- efficient logistics,
- minimal excess inventory.

Economic growth aligns with environmental preservation.

Example - Circular Economy Platforms

AI tracks product lifecycle:

- reuse opportunities identified,

- recycling automated,
- materials continuously reused.

Consumption becomes regenerative rather than extractive.

ZHA Insight

Feeling Sufficient replaces growth driven by excess with growth driven by efficiency and balance.

13.12.6 SDG 13 - Sustainable Cities and Communities

AI as Urban Civilization Infrastructure

Urbanization is accelerating globally.

Cities face challenges:

- congestion,
- pollution,
- infrastructure strain,
- social fragmentation.

AI enables intelligent urban ecosystems.

Responsible AI Application

Example - Smart Mobility Systems

AI traffic optimization:

- reduces emissions,
- minimizes commute times,
- improves urban quality of life.

Example - AI Energy Management
Buildings automatically adjust energy consumption based on usage patterns.

Cities become sustainable ecosystems.

ZHA Alignment

Sustainable cities reflect the ZHA pillars:

- People → inclusive communities
- Planet → environmental protection
- Prosperity → economic vitality
- Partnership → collaborative governance

Cities evolve into living sustainability systems.

13.12.7 From SDGs to Sustainable AI Civilization

The UN SDGs provide humanity's destination.

The ZHA Sustainability Mindset provides the operating model to reach it.

When AI transformation aligns with SDGs:

- innovation serves humanity,
- technology strengthens ecosystems,
- economic growth becomes sustainable.

AI becomes civilization infrastructure supporting global harmony.

13.12.8 The ZHA Insight - SDGs as Intelligence Objectives

Traditional organizations treat SDGs as external commitments.

ZHA reframes SDGs as design objectives for intelligent systems.

AI systems are designed to optimize:

- education access,
- climate resilience,
- social inclusion,
- responsible consumption,
- sustainable urban living.

Technology becomes aligned with human progress by design.

13.12.9 Responsible Innovation as Global Collaboration

Sustainable development cannot be achieved by one country or one organization alone.

Responsible innovation requires collaboration between:

- governments,
- academia,
- industry,
- communities,
- global institutions.

AI becomes a shared human project.

Thus, When innovation aligns with the UN Sustainable Development Goals:

Artificial Intelligence stops being a technological race. It becomes a civilization mission.

The future belongs to innovators who build intelligence that advances humanity, protects the planet, and enables sustainable prosperity for generations to come.

13.13 The Role of ZHA in Global Sustainable AI Governance

Building the Human-Centric Intelligence Ecosystem

13.13.1 From Framework to Global Movement

Artificial Intelligence transformation is often approached as a technological upgrade.

Organizations invest in:

- automation tools,
- machine learning platforms,
- generative AI capabilities,
- data infrastructure modernization.

Yet most AI transformations fail to achieve sustainable outcomes because they focus primarily on technology adoption rather than human transformation.

The ZHA Sustainability Mindset framework addresses this missing dimension.

ZHA introduces AI transformation as:

- a civilization-scale evolution integrating mindset,
- technology,
- governance,
- societal responsibility.

ZHA therefore functions not only as a framework but as a global coordination model guiding how humanity adopts intelligent systems responsibly.

13.13.2 The Core Mission of ZHA in AI Transformation

ZHA enables Sustainable AI Governance by aligning four transformation layers:

This multi-layer integration ensures AI transformation does not destabilize social systems.

13.13.3 Educating Students - Creating Responsible Innovators

ZHA begins transformation at the earliest stage of civilization development: education.

Traditional education prepares students for employment. ZHA prepares students for civilization responsibility.

Through initiatives such as ZSPC, students learn to:

- understand sustainability challenges,
- evaluate technological consequences,
- collaborate on real-world problems,
- develop ethical awareness alongside technical skills.

Example - ZSPC Sustainability Practitioner Community Model

Students participate in:

- environmental restoration activities,
- community sustainability projects,
- ethical technology discussions,
- innovation challenges addressing local problems.

AI transformation impact:

Future engineers enter industry already aligned with ESG thinking. Governance becomes preventive culture, not reactive regulation.

13.13.4 Guiding Professionals - Transforming the Workforce

Professionals represent the operational backbone of AI transformation.

Without mindset evolution, professionals may unintentionally deploy technologies that create long-term harm.

ZHA guides professionals to evolve into Responsible AI Practitioners.

Professionals learn to integrate:

- ethical decision-making,
- sustainability evaluation,
- human-centered design thinking,
- risk-aware innovation practices.

Example - Enterprise AI Team Transformation

A software development organization adopting ZHA principles introduces:

- ESG checkpoints within SDLC,
- ethical review during feature design,
- energy efficiency evaluation during model training.

Result: Innovation speed remains high while risk decreases.

Professionals transition from system builders to technology stewards.

13.13.5 Supporting Organizations - Enabling Sustainable AI Transformation

Organizations face intense pressure to adopt AI rapidly to remain competitive.

However, unmanaged adoption leads to:

- workforce disruption,
- ethical failures,

- environmental cost escalation,
- loss of public trust.

ZHA provides organizations with a structured transformation pathway.

ZHA Organizational Enablement Model

Organizations integrate:

- Sustainability Mindset culture.
- ESG-aligned AI governance.
- Responsible SDLC methodologies.
- Human-centered product development.

Example - Responsible AI Enterprise

An organization implementing ZHA transformation:

- automates operations responsibly,
- retrains employees instead of replacing them,
- deploys explainable AI models,
- monitors ESG impact continuously.

Outcome: AI transformation strengthens both economic performance and societal trust.

13.13.6 Influencing Governance Structures — From Policy to Culture

Global AI governance cannot rely solely on legislation.
Regulation often reacts after technological harm occurs.

ZHA introduces culture-driven governance.

Through collaboration with:

- educational institutions,
- industries,
- communities,
- policy stakeholders,

ZHA influences governance structures indirectly by shaping the people who design and operate intelligent systems.

Example - Policy Influence Through Practice

When universities adopt sustainability mindset education:

- graduates carry ethical awareness into industry,
- organizations naturally improve governance,
- policymakers observe improved outcomes and adopt supportive frameworks.

Governance evolves organically.

13.13.7 ZHA as the Bridge Between Education, Innovation, and Civilization

Modern civilization faces fragmentation:

- education disconnected from industry,
- innovation disconnected from ethics,
- governance disconnected from society.

ZHA reconnects these domains.

The ZHA Bridge Model
Education → Innovation → Governance → Civilization

Education builds mindset.
Innovation applies intelligence.

Governance ensures responsibility.
Civilization achieves sustainability.

ZHA becomes the connecting architecture enabling alignment.

13.13.8 ZHA and the AI Transformation Lifecycle

ZHA integrates across the entire AI transformation journey:

Align
Define purpose before technology adoption.

Automate
Introduce AI responsibly.

Augment
Enhance human capability.

Innovate
Create human-centered solutions.

Sustain & Scale
Ensure long-term societal benefit.

ZHA ensures AI transformation becomes evolutionary rather than disruptive.

13.13.9 Scaling Responsible Innovation Globally

The ZHA model is inherently scalable because it operates through mindset rather than infrastructure alone.

It can be adopted by:

- schools,
- universities,
- corporations,
- governments,

- communities.

This scalability enables global adoption of Sustainable AI Governance without requiring identical political or economic systems.

Example - Global Sustainability Practitioner Network

Students, professionals, and organizations across regions participate in shared sustainability practices and certifications.

Innovation becomes globally collaborative rather than competitively destructive.

13.13.10 The ZHA Insight - From AI Adoption to Civilization Stewardship

Most AI strategies aim for digital transformation.

ZHA advances toward Civilization Transformation.

Artificial Intelligence becomes:

- ethically guided,
- environmentally responsible,
- socially inclusive,
- governance-aligned.

ZHA positions humanity to transition from technological acceleration toward conscious evolution.

Thus, ZHA is not merely an organization.

It is a human-centered intelligence ecosystem.

By educating students, guiding professionals, supporting organizations, and influencing governance structures, ZHA becomes the bridge connecting innovation with civilization stewardship.

In the age of Artificial Intelligence, the greatest innovation is not smarter machines — but wiser humanity.

13.14 Responsible Innovation as Civilization Evolution

Human civilization advances not through intelligence alone but through responsible intelligence.

13.14.1 Intelligence Alone Does Not Advance Civilization

Throughout history, humanity has often equated intelligence with progress.

Scientific discovery, industrial capability, and technological innovation have transformed human life dramatically.

Yet history reveals a deeper truth:

Civilizations do not advance merely because they become more intelligent. They advance when intelligence becomes responsible.

The same intelligence that produces medicine can create weapons.
The same technology that connects humanity can divide societies.
The same innovation that increases productivity can damage ecosystems.

Intelligence amplifies human intention.

Therefore, civilization evolution depends not on intelligence alone, but on responsible intelligence.

13.14.2 The AI Era - Humanity's Defining Moment

Artificial Intelligence represents the first technology capable of scaling decision-making itself.

For the first time:

- machines influence human thought,

- algorithms shape economic behavior,
- intelligent systems guide governance decisions.

AI does not merely extend human capability — it reshapes human civilization.

This creates an unprecedented responsibility.

The AI era demands innovators who understand a fundamental chain of influence:

- Technology shapes society.
- Society shapes civilization.
- Civilization determines humanity's future.

Every line of code now participates in civilization design.

13.14.3 Technology Shapes Society

Technology alters how humans live, work, and relate to one another.

Example - Social Media Algorithms

Algorithms optimized for engagement transformed communication:

- Positive outcomes:
- global connectivity,
- information accessibility,
- community formation.

Unintended consequences:

- polarization,
- misinformation,
- psychological dependency.

Technology reshaped societal behavior faster than governance evolved.

Responsible innovation requires anticipating these societal effects before deployment.

Example - AI Automation

Automation increases efficiency but also reshapes employment structures.

Responsible innovators ask:

- How do workers transition?
- What new opportunities are created?
- How do we preserve dignity alongside productivity?

Technology must enhance human flourishing, not undermine it.

13.14.4 Society Shapes Civilization

Societal values determine how technologies influence civilization.

If society values:

- competition without ethics → instability grows.
- consumption without restraint → environmental collapse accelerates.
- innovation without responsibility → trust erodes.

Conversely, societies grounded in sustainability produce resilient civilizations.

Example - Renewable Energy Transition

Societies choosing sustainability invest in:

- clean energy,

- circular economies,
- ecological restoration.

Technology aligned with societal values transforms civilization toward regeneration rather than extraction.

ZHA emphasizes that civilization outcomes are reflections of collective mindset.

13.14.5 Civilization Determines Humanity's Future

Civilization is humanity's long-term operating system.

It determines:

- environmental survival,
- social harmony,
- economic stability,
- intergenerational wellbeing.

Artificial Intelligence will define whether future civilization becomes:

- hyper-efficient but unstable, or
- intelligent and sustainable.

Responsible innovation ensures civilization evolves toward balance rather than collapse.

Example - AI in Governance

AI-assisted policymaking can:

- predict social risk,
- optimize public resource allocation,
- prevent crises before escalation.

When guided responsibly, AI strengthens democratic resilience.

13.14.6 Responsible Innovation as Evolutionary Intelligence

ZHA introduces a new concept:

Evolutionary Intelligence.

Human progress evolves through stages:

Responsible innovation represents humanity's transition from technological maturity to civilizational maturity.

13.14.7 The Role of Innovators in Civilization Evolution

In previous eras, innovators built products.

In the AI era, innovators shape humanity's trajectory.

Modern innovators must therefore become:

- ethical thinkers,
- sustainability advocates,
- governance participants,
- civilization designers.

Example - AI Healthcare Innovator

A responsible innovator ensures:

- algorithms reduce healthcare inequality,
- systems remain explainable,
- patient dignity remains central.
- Innovation becomes public service.

13.14.8 Sustainable AI - The Foundation of Humane Civilization

Responsible innovation ensures the future remains:

Sustainable
Technology protects planetary resources.

Inclusive
Innovation benefits all communities.

Humane
Human dignity remains central to intelligence systems.

Under the ZHA Sustainability Mindset, AI becomes aligned with human values rather than replacing them.

13.14.9 The ZHA Insight - Civilization Is a Design Choice

Civilizations do not collapse because technology exists.

They collapse when innovation lacks responsibility.

Humanity now possesses intelligence powerful enough to transform the planet.

The question is no longer:

1. Can we innovate?
2. The question becomes:
3. Can we innovate wisely?

ZHA answers this by positioning responsible innovation as civilization evolution itself.

13.14.10 From Innovation to Human Destiny

Responsible innovation transforms Artificial Intelligence from a competitive race into a shared human mission.

Innovation guided by sustainability creates:

- resilient economies,
- harmonious societies,
- protected ecosystems,
- enduring human progress.

The AI era therefore marks humanity's transition from technological civilization to conscious civilization.

Thus, Human civilization advances not through intelligence alone but through responsible intelligence.

Technology shapes society.
Society shapes civilization.
Civilization determines humanity's future.

Responsible innovation ensures this future remains sustainable, inclusive, and humane.

The innovators of the AI age are no longer merely creators of technology — they are architects of humanity's destiny.

PART IV — Building the ZHA AI Transformation Global Ecosystem

CHAPTER XIV

Chapter 14: ZHA Global Institute

Introduces institutional structures designed to scale Sustainable AI globally through governance, research, and collaborative leadership.

14.1 Why Humanity Needs Global Sustainable AI-Aligned Institutions and Organizations

Institutional Intelligence for the Age of Artificial Intelligence

14.1.1 The Institutional Gap of the AI Era

Human civilization has always progressed through institutions.

Institutions stabilize transformation.

Universities stabilized scientific discovery.

Governments stabilized political systems.

Financial institutions stabilized economic exchange.

International organizations stabilized globalization.

Every major technological revolution succeeded because institutions evolved alongside technology.

Today humanity faces a unique challenge:
Artificial Intelligence is advancing faster than institutional evolution itself.

For the first time in history, technology is no longer waiting for governance, education, or ethics to mature. AI innovation moves at exponential speed while institutional adaptation remains linear. This mismatch creates a profound institutional gap.

14.1.2 A World Where Technology Is Global but Institutions Are Not

Artificial Intelligence operates without borders.

Algorithms developed in one country immediately influence:

- global markets,
- education systems,
- democratic processes,
- environmental outcomes,
- cultural narratives.

Yet humanity governs AI through fragmented systems:

- Current Reality
- Technologies evolve globally.
- Governance remains nationally confined.
- Education systems remain disconnected.

Ethical standards vary widely across organizations.

Example - Generative AI Deployment

A generative AI system released globally affects:

- students in developing nations,
- journalists worldwide,
- businesses across continents.

However:

- regulations differ country to country,
- ethical guidelines remain inconsistent,
- education systems are unprepared for AI literacy.

Technology becomes global faster than responsibility.

14.1.3 Fragmented Education in an Intelligent World

Education remains one of the most fragmented global systems.

Most educational institutions still prepare students for:

- industrial-era jobs,
- static knowledge models,
- localized career paths.

But AI requires:

- lifelong learning,
- ethical reasoning,
- interdisciplinary thinking,
- sustainability awareness.

Without institutional transformation, societies risk producing technologically capable individuals without responsible mindset formation.

Example - AI-Skilled but Ethically Unprepared Workforce

A developer trained only in coding efficiency may build:

- surveillance algorithms,
- manipulative recommendation systems,
- resource-intensive AI models.

The issue is not lack of intelligence - but lack of institutional guidance toward responsibility.

ZHA identifies this as a mindset deficit, not a talent deficit.

14.1.4 Governance Lag in the AI Revolution

Traditional governance structures were designed for slower technological change. Policy cycles operate over years, while AI innovation evolves in months or even weeks.

As a result:

- regulations arrive after harm occurs,
- governance becomes reactive,
- public trust declines.

Example - Social Media Governance

Social platforms scaled globally before societies understood:

- misinformation dynamics,
- mental health impacts,
- algorithmic polarization.

Governance followed disruption rather than preventing it.

AI transformation risks repeating this pattern unless institutions evolve.

14.1.5 Inconsistent Ethical Standards Across Organizations

Today, ethical AI depends largely on organizational choice.

Some organizations adopt strong ESG practices.

Others prioritize speed and market dominance.

This inconsistency produces uneven outcomes:

- ethical innovation in some regions,

- technological exploitation in others.
- Humanity lacks a shared institutional compass guiding intelligent systems.

Example - AI Hiring Systems

One company audits bias carefully. Another deploys untested models. Both technologies influence livelihoods, yet ethical responsibility varies dramatically.

Without aligned institutions, civilization experiences uneven justice.

14.1.6 Reactive ESG Protection in the AI Age

Environmental, Social, and Governance (ESG) frameworks were introduced to improve corporate responsibility.

However, many ESG practices remain retrospective:

- annual sustainability reports,
- compliance documentation,
- reputation management initiatives.

AI changes the nature of risk.

Because intelligent systems continuously evolve, ESG protection must also become continuous.

Reactive ESG is insufficient for adaptive intelligence.

Example - Energy Consumption of AI Models

Large-scale AI training consumes enormous energy resources.

Organizations often measure impact after deployment rather than designing sustainability into model architecture.

ZHA introduces ESG as embedded operational intelligence rather than external reporting.

14.1.7 Innovation Without Civilizational Coordination

Human innovation today operates primarily through competition.

Companies race to release more powerful AI systems.

Nations compete for technological dominance.

Researchers pursue breakthroughs independently.

While competition accelerates innovation, it lacks coordination toward shared human outcomes.

This produces technological acceleration without civilizational alignment.

Example - Autonomous Weapons Risk

AI advancements enable autonomous defense systems globally.

Without coordinated institutions:

- escalation risks increase,
- ethical consensus weakens,
- global stability becomes fragile.

Civilization requires cooperative intelligence alongside competitive innovation.

14.1.8 The Emergence of Institutional Intelligence

Humanity has entered a new phase of evolution:

Technological intelligence now exceeds institutional intelligence.

To stabilize the AI era, humanity must develop Institutional Intelligence — organizations capable of guiding technological power responsibly.

Institutional intelligence integrates:

- education,
- governance,
- ethics,
- innovation,
- sustainability.

It ensures humanity evolves consciously alongside its creations.

14.1.9 The ZHA Global Institute - A Civilizational Response

The ZHA Global Institute emerges as a response to this historical moment.

It is designed not merely as:

- a research center,
- a training institution,
- or a policy organization.

It represents a new category of institution:
A Civilization Guidance Institution for the Age of Artificial Intelligence.

Its purpose is to align:

- intelligence with humanity,
- innovation with sustainability,
- governance with long-term civilization wellbeing.

14.1.10 Guiding Humanity's Relationship with Intelligence

Humanity's greatest challenge is no longer technological capability.

It is defining how humans coexist with intelligent systems.

The ZHA Global Institute addresses this by guiding humanity's relationship with intelligence itself.

It helps societies answer fundamental questions:

- How should intelligence serve humanity?
- What responsibilities accompany technological power?
- How can innovation preserve planetary balance?
- How can AI strengthen rather than replace human purpose?

Example - National AI Transformation Partnership

A nation collaborates with the ZHA Global Institute to:

- integrate sustainability mindset education,
- establish ESG-aligned AI governance,
- support ethical industry innovation,
- promote human-centered digital transformation.

AI adoption becomes civilization advancement rather than disruption.

14.1.11 Toward a Coordinated Intelligent Civilization

The future of humanity depends on coordinated action across:

- education systems,
- industries,
- governments,
- communities.

Global Sustainable AI-aligned institutions provide the structure required to harmonize these forces.

They transform technological progress into civilizational progress.

Insight

Artificial Intelligence does not merely transform industries. It transforms civilization itself. Humanity now requires institutions capable of guiding intelligence responsibly.

The ZHA Global Institute emerges as one such institution — not to control technology, but to ensure intelligence evolves in harmony with humanity, society, and the planet

14.2 From AI Adoption to AI Civilization Governance

Transforming Artificial Intelligence into Civilization Infrastructure

14.2.1 The Current Understanding of AI Transformation

Across industries, Artificial Intelligence is largely approached as a technological upgrade.

Organizations initiate AI transformation through:

- automation programs replacing manual work,
- productivity optimization initiatives,
- digital modernization projects,
- cost reduction strategies,
- competitive innovation races.

These efforts are valuable but limited.

They treat AI as a tool for organizational efficiency, not as a force reshaping civilization itself.

As a result, AI adoption often focuses on:

- speed,
- scalability,
- profit,
- operational performance.

What remains missing is civilizational intent.

Example - Enterprise Automation

A corporation deploys AI to automate customer service.

Outcome:

- operational costs decrease,
- response time improves.

But unaddressed questions remain:

- How are displaced employees supported?
- Does automation increase inequality?
- Are customers interacting with transparent systems?
- What long-term societal behavior does this create?

AI adoption occurs - but civilization impact remains unmanaged.

14.2.2 The Limitation of Technology-Centric AI Adoption

Technology-centered transformation creates three systemic risks:

1. Fragmented Innovation
Organizations innovate independently without shared responsibility.

2. Governance Lag
Regulation follows disruption rather than guiding it.

3. Trust Erosion
Societies begin to distrust intelligent systems.

When AI grows without coordinated governance, innovation accelerates while social stability weakens.

Humanity risks technological advancement without civilizational maturity.

14.2.3 ZHA Reframing - AI as Civilization Infrastructure

ZHA introduces a transformative perspective:

AI is not merely technology.
Sustainable AI is Civilization Infrastructure.

Just as electricity, transportation, and the internet became foundational societal systems, Artificial Intelligence now forms the infrastructure governing:

- knowledge creation,
- economic decision-making,
- healthcare delivery,
- education systems,
- governance processes.

AI therefore requires institutional stewardship comparable to public infrastructure.

Example - AI in Healthcare Systems

AI diagnostics influence millions of medical decisions.

If treated as software:
→ deployed commercially.

If treated as civilization infrastructure:
→ governed ethically,

→ monitored continuously,
→ aligned with societal wellbeing.

The difference is governance philosophy.

14.2.4 The Three Institutional Shifts Introduced by ZHA

The ZHA Global Institute institutionalizes a fundamental transition:
From Artificial Intelligence Adoption → Sustainable AI Civilization Governance

Artificial Intelligence is no longer merely a technological capability. It has become an organizing force shaping economies, societies, governance systems, and human behavior.

To guide this transformation responsibly, ZHA introduces three institutional shifts that redefine how humanity relates to intelligent systems.

Shift 1 - AI as Technology → Sustainable AI as Civilization Infrastructure

Traditional Model

Artificial Intelligence is viewed as:

- software capability,
- automation solution,
- productivity enhancer,
- competitive advantage.

Organizations deploy AI primarily to optimize efficiency and market position.

AI remains an internal business tool.

ZHA Sustainable AI Model

Artificial Intelligence becomes:

- a societal decision system,
- public trust infrastructure,
- sustainability enabler,
- civilization stabilizer.

Under ZHA alignment, AI is treated similarly to electricity, healthcare, or transportation - critical infrastructure supporting civilization itself.

Sustainable AI ensures intelligence serves:

- people,
- planet,
- prosperity,
- future generations.

Example - Search Algorithms as Civilization Infrastructure

Search engines now influence:

- public knowledge,
- democratic discourse,
- education access,
- societal perception of truth.

Under ZHA Sustainable AI Governance:

- misinformation risks are continuously assessed,
- algorithmic impact on society is monitored,
- transparency mechanisms are embedded,
- ethical accountability frameworks operate permanently.

Technology becomes public responsibility, not private power.

Sustainable AI Outcome

AI systems are designed for:

- long-term societal stability
- environmental sustainability
- inclusive knowledge access

AI evolves into civilization infrastructure.

Shift 2 - Governance as Regulation → Governance as Stewardship

Traditional Governance

Governments and institutions regulate after problems occur.

Governance becomes reactive:

- laws follow disruption,
- compliance responds to crisis,
- innovation and ethics remain disconnected.

Regulation attempts to control technology after damage appears.

ZHA Stewardship Governance Model

ZHA reframes governance as Stewardship.

Organizations, educators, policymakers, and innovators become guardians of intelligent systems.

Governance becomes:

- anticipatory,
- collaborative,

- value-driven,
- sustainability-centered.

Stewardship means responsibility exists before deployment, not after failure.

Example - Autonomous Vehicles

Regulatory Governance

- approve safety compliance,
- allow deployment,
- respond to accidents.

Sustainable AI Stewardship

- evaluate environmental impact,
- prepare workforce transition,
- ensure accessibility equity,
- assess urban infrastructure implications,
- build citizen trust frameworks.

Governance guides innovation toward human wellbeing rather than constraining creativity.

Sustainable AI Outcome

Governance evolves from control → care.

AI becomes trusted because society participates in its stewardship.

Shift 3 - Innovation as Competition → Innovation as Collective Progress

Traditional Innovation Model

Organizations compete aggressively for AI dominance.

Consequences include:

- duplicated research effort,
- fragmented ethical standards,
- uneven societal benefits,
- accelerated technological inequality.

Innovation advances capability but not civilization harmony.

ZHA Collective Progress Model

Sustainable AI requires collective intelligence.

Innovation becomes collaborative civilization advancement.

Organizations cooperate by sharing:

- sustainability standards,
- ethical design practices,
- ESG methodologies,
- responsible research insights.

Innovation serves humanity collectively rather than competitively.

Example - Climate AI Collaboration

Instead of proprietary competition:

Multiple companies, universities, and governments collaborate on AI climate models to:

- predict environmental risks,
- optimize renewable energy,
- reduce global emissions.

Sustainable AI transforms innovation into planetary stewardship.

Sustainable AI Outcome

Progress becomes shared human advancement rather than technological rivalry.

4.2.5 Global AI Deployment - The Coordination Problem

Artificial Intelligence operates globally the moment it is released.

However, deployment today remains decentralized.

Without Coordination:

- organizations deploy AI independently,
- ethical standards vary widely,
- ESG protections differ across regions,
- societal impacts become unpredictable,
- public trust declines.

Example - Facial Recognition Systems

Different countries deploy systems with inconsistent:

- privacy protections,
- bias controls,
- governance transparency.

Human rights protection becomes uneven.

Civilization experiences technological fragmentation.

Sustainable AI cannot emerge without coordination.

14.2.6 The ZHA Global Institute as Coordinating Intelligence

The ZHA Global Institute introduces institutional coordination for Sustainable AI.

It functions as humanity's coordinating intelligence layer.

The Institute enables:

- shared Sustainable AI governance frameworks,
- globally aligned ESG standards,
- collaborative research ecosystems,
- cross-sector leadership platforms,
- education–industry–policy integration.

Instead of isolated innovation, humanity develops coordinated intelligence systems.

Example - Global AI Deployment with ZHA Alignment

Through Institute collaboration:

- organizations adopt shared ESG metrics,
- universities integrate Responsible AI curricula,
- governments receive sustainability-aligned policy guidance,
- industries coordinate ethical innovation practices.

Outcome

AI transformation becomes:

- globally aligned
- sustainability-driven
- trust-centered
- civilization-oriented

rather than competitively fragmented.

14.2.7 AI Transformation Becomes Civilization Governance

Under ZHA alignment, AI transformation evolves through three stages.

Organizations transition from:
technology users → civilization participants.

AI governance becomes distributed across:

- institutions,
- education systems,
- industries,
- communities.

Responsibility becomes shared intelligence.

14.2.8 Real-World Scenario - National Sustainable AI Transformation

A nation adopting ZHA Civilization Governance implements:

- Sustainability mindset education integrated into schools.
- ESG-aligned AI adoption across industry.
- Ethical research collaboration among universities.
- Transparent Sustainable AI governance frameworks.

Results

- economic innovation accelerates,
- social inclusion improves,
- environmental protection strengthens,
- citizen trust increases.

AI becomes a stabilizing national infrastructure.

14.2.9 The ZHA Insight — The Future Belongs to Governed Sustainable Intelligence

The greatest risk of Artificial Intelligence is not intelligence itself.

It is intelligence without coordination and sustainability.

Civilizations thrive when infrastructure is responsibly governed.

Humanity previously institutionalized governance for:

- transportation,
- finance,
- healthcare,
- energy systems.

Artificial Intelligence now requires equivalent institutional maturity.

The ZHA Global Institute represents humanity's transition toward: Governed Sustainable Intelligence

Insights

AI adoption changes organizations.

Sustainable AI civilization governance changes humanity.

The future will not belong to those who build the most powerful AI systems.

It will belong to those who build the most responsible, sustainable, and human-centric intelligence ecosystems.

Through the ZHA Global Institute, AI transformation evolves from technological disruption into coordinated civilization progress.

14.3 The Mission of the ZHA Global Institute

The Institute exists to scale:

- Sustainable AI
- Responsible Innovation
- ESG-aligned AI Transformation
- Human-Centric Intelligence Leadership

Its mission:
To align intelligence, institutions, and humanity toward sustainable civilization evolution.

14.4 Core Functions of the ZHA Global Institute

The Institute operates across four integrated pillars.

1. Governance Leadership
Develop global frameworks for Sustainable AI governance.

Activities include:

- AI ESG standards development,
- ethical policy advisory,
- institutional governance models,
- cross-country collaboration forums.

Example

Governments consult the Institute while drafting AI policy ensuring:

- environmental impact consideration,
- workforce transition planning,
- ethical algorithm deployment.

Governance becomes proactive rather than crisis-driven.

2. Research & Knowledge Creation
The Institute functions as a global research hub integrating:

- AI engineering,
- sustainability science,
- behavioral psychology,
- social governance.

Example

Research teams study:

- carbon footprint of AI models,
- human-AI collaboration effectiveness,
- ethical design frameworks.

Outputs influence global innovation practices.

3. Education & Capacity Building
ZHA institutionalizes Sustainability Mindset education worldwide.

Programs include:

- ZSPC student certifications,
- professional Sustainable AI certifications,
- executive leadership programs,
- governance fellowships.

Example - AI Transformation Leadership Program

Executives learn:

- ESG-aligned AI adoption,
- human-centered product design,
- long-term civilization risk thinking.

Leaders become AI stewards instead of technology adopters.

4. Collaborative Leadership Platform

The Institute connects:

- universities,
- corporations,
- governments,
- NGOs,
- communities.

Innovation becomes collective intelligence.

Example - Global Climate AI Collaboration

Multiple organizations jointly develop AI solutions for climate resilience rather than competing independently.

14.5 AI Transformation with ZHA Sustainability Mindset Alignment

The ZHA Institute operationalizes AI transformation through the ZHA Alignment Model.

Step 1 - Align (Mindset First)

Before adopting AI:

Organizations define:

- purpose,
- sustainability values,

- societal impact goals.

Example
A manufacturing firm defines carbon reduction goals before automation rollout.

AI adoption begins with intention.

Step 2 - Automate Responsibly

Automation introduced where it enhances human productivity.

Example
AI handles repetitive administrative tasks while employees transition into analytical and creative roles.

Human dignity preserved.

Step 3 - Augment Human Capability

AI expands human intelligence rather than replacing it.

Example
Doctors use AI diagnostics as decision support, improving accuracy while retaining human judgment.

Step 4 — Innovate Sustainably

Organizations develop new human-centered solutions.

Example
AI-enabled agriculture increases yield while conserving water resources.

Innovation supports planet and prosperity simultaneously.

Step 5 — Sustain & Scale

Long-term ESG monitoring ensures transformation remains responsible.

Example
Continuous AI auditing tracks:

- energy usage,
- workforce impact,
- societal trust indicators.

Transformation becomes stable evolution.

14.6 Institutional Architecture of the ZHA Global Institute

The Institute operates through layered global structures:

Global Council
Strategic civilization guidance.

Research Think Tank
Knowledge creation and foresight analysis.

Education Network
Student and professional development ecosystem.

Industry Consortium
Corporate AI transformation collaboration.

Policy Advisory Forum
Government and institutional engagement.

Together, these form a Global Sustainable AI Governance System.

14.7 The ZHA Global Institute as Civilization Stabilizer

Historically, institutions stabilized previous revolutions:

- Universities stabilized scientific revolutions.

- International organizations stabilized economic globalization.
- The ZHA Global Institute stabilizes the AI Revolution.

Its role is to ensure:

- innovation does not outpace ethics,
- intelligence does not outgrow responsibility,
- progress does not threaten humanity.

14.8 Real-World Transformation Scenario

Scenario: National AI Transformation

A country partners with the ZHA Global Institute.

Implementation includes:

- sustainability mindset education in schools,
- ESG-aligned corporate AI adoption,
- ethical governance frameworks,
- AI research collaboration.

Outcomes:

- economic growth,
- social inclusion,
- environmental protection,
- global innovation leadership.

AI transformation strengthens national civilization resilience.

14.9 ZHA Sustainable AI as the Bridge Between Innovation and Sustainable AI Civilization

Integrating Intelligence, Institutions, and Humanity

14.9.1 The Fragmentation of the Modern World

Modern civilization has achieved extraordinary technological advancement. However, progress has evolved in disconnected domains:

- education develops skills but not responsibility,
- industry drives innovation but often ignores sustainability,
- governance reacts to disruption rather than guiding it,
- society consumes technology without participating in its direction.

Each domain advances independently.

Artificial Intelligence magnifies this fragmentation.

Technology connects systems digitally while institutions remain philosophically disconnected.

This produces a paradox:

Humanity is technologically connected but civilizationally fragmented.

The ZHA Global Institute addresses this fragmentation by acting as a bridge institution, aligning innovation with Sustainable AI civilization.

14.9.2 Why Civilization Requires Integration

Civilization stability depends on alignment among four forces:

- Knowledge Creation
- Economic Innovation
- Governance Systems
- Social Participation

When these forces operate independently:

- innovation outpaces ethics,
- governance becomes reactive,
- society loses trust,
- sustainability becomes secondary.

Sustainable AI requires coordinated evolution across all domains simultaneously.

ZHA introduces this coordination.

14.9.3 The ZHA Integration Model

The ZHA Global Institute connects previously disconnected domains through Sustainable AI principles.

ZHA transforms fragmented systems into an integrated civilization ecosystem.

14.9.4 Education - From Skills to Sustainability Mindset

Traditional Gap

Education systems historically focused on:

- technical competency,
- academic achievement,
- employment readiness.

Students graduate with knowledge but limited understanding of technology's societal consequences.

ZHA Sustainable AI Integration

ZHA embeds Sustainability Mindset education:

Students learn to ask:

- What problem does this innovation solve for humanity?
- What environmental impact does this technology create?
- How does intelligence serve society?

Example - ZSPC Education Ecosystem

Students participating in ZSPC programs:

- conduct sustainability projects,
- analyze AI ethical scenarios,
- collaborate on community solutions,
- understand governance responsibilities early.

Outcome:

Education produces Responsible Innovators, not only skilled workers.

AI civilization begins in classrooms.

14.9.5 Industry - From Profit Focus to Sustainable Innovation

Traditional Gap

Industry innovation historically prioritizes:

- revenue growth,
- market dominance,
- rapid scaling.

While economically effective, this approach often overlooks:

- environmental cost,
- workforce impact,
- long-term societal consequences.

ZHA Sustainable AI Integration

ZHA reframes innovation through Sustainable AI principles:

Organizations evaluate success using:

- ESG outcomes,
- societal value creation,
- environmental responsibility,
- human wellbeing indicators.

Example - Sustainable AI Product Development

A technology company deploying AI customer analytics:

Traditional approach:
maximize engagement regardless of psychological impact.

ZHA approach:

- implement ethical design,
- prevent manipulative algorithms,
- ensure transparency,
- measure wellbeing outcomes.

Profitability aligns with sustainability.

Industry evolves into innovation stewardship.

14.9.6 Governance - From Reactive Regulation to Anticipatory Stewardship

Traditional Gap

Governments and institutions typically regulate after disruption occurs.

Policy follows innovation rather than guiding it.

ZHA Sustainable AI Integration

Governance becomes anticipatory.

Through ZHA alignment:

- policymakers collaborate with educators and industry,
- sustainability risks assessed before deployment,
- AI ESG frameworks embedded into development cycles.

Example - AI Public Infrastructure

Before implementing AI-based public surveillance:

ZHA governance requires:

- community consultation,
- ethical risk evaluation,
- transparency protocols,
- long-term societal assessment.

Governance shifts from control to stewardship.

14.9.7 Society - From Passive Consumption to Participatory Civilization

Traditional Gap

Society traditionally acts as a technology consumer.

Citizens rarely participate in shaping technological direction.

This creates distrust and resistance.

ZHA Sustainable AI Integration

ZHA enables participatory innovation.

Citizens become:

- contributors,
- evaluators,
- collaborators in AI transformation.

Example - Community Sustainability Platforms

Local communities collaborate with AI researchers to:

- monitor environmental health,
- design urban sustainability solutions,
- participate in governance discussions.

Technology becomes socially owned rather than externally imposed.

14.9.8 ZHA as the Bridge Institution

The ZHA Global Institute connects:

- education with innovation,
- innovation with governance,
- governance with society,
- society with civilization sustainability.

ZHA functions as a Civilization Bridge Institution.

It ensures intelligence evolves alongside:

- ethical maturity,
- environmental responsibility,

- social inclusion,
- long-term human purpose.

The ZHA Bridge Model
Mindset → Innovation → Governance → Civilization

ZHA synchronizes these layers into a coherent Sustainable AI ecosystem.

14.9.9 Sustainable AI Civilization - The Emerging Outcome

When domains align under ZHA principles:

- education produces responsible innovators,
- industry builds sustainable technologies,
- governance anticipates disruption,
- society participates in decision-making.

Artificial Intelligence transforms from disruptive force into civilization stabilizer.

Sustainable AI becomes the operating foundation of future civilization.

14.9.10 The ZHA Insight - Intelligence Requires Integration

The greatest challenge of the AI age is not technological capability.

It is integration.

Technology alone cannot create sustainable civilization.

Only when innovation, governance, education, and society align can intelligence serve humanity responsibly.

ZHA provides this missing integration layer.

Insights

ZHA is not simply an institution. It is the bridge linking intelligence with civilization stewardship.

Through Sustainable AI with Human-Centric Purpose, ZHA connects fragmented progress into unified human advancement — guiding humanity toward a sustainable, intelligent, and conscious civilization.

14.10 The ZHA Insight - Institutions & Organizations Shape the Future of Intelligence

Why Sustainable AI Requires Civilizational Institutions

14.10.1 Every Technological Revolution Required Institutions

Human history reveals a consistent and powerful pattern:

Technological revolutions succeed not because of inventions alone, but because societies build institutions capable of guiding those inventions responsibly.

Technology creates possibility. Institutions create stability.

Civilization progresses when innovation and institutional wisdom evolve together.

Historical Perspective

The Agricultural Revolution

Agriculture transformed human survival, enabling settlement and population growth.
However, civilization emerged only when societies created:

- governance systems,
- community cooperation structures,
- shared resource management practices.

Farming created abundance; institutions created civilization.

The Industrial Revolution

Machines dramatically increased productivity and economic expansion.

Without institutional evolution such as:

- labor protection laws,
- public education systems,
- economic governance frameworks,
- industrialization would have produced widespread instability rather than prosperity.

Institutions transformed industrial power into societal progress.

The Digital Revolution

The internet connected humanity globally.

Its success depended on institutions that ensured coordination and trust:

- global standards organizations,
- cybersecurity governance systems,
- international cooperation frameworks.

Technology advanced because institutions matured alongside it.

14.10.2 Artificial Intelligence - A Different Kind of Revolution

Artificial Intelligence represents a fundamentally new category of technological transformation.

Previous technologies amplified human physical capability.

Artificial Intelligence amplifies human cognition.

AI now influences:

- how decisions are made,
- how knowledge is created,
- how economies function,
- how societies organize governance.

For the first time in history, humanity is building systems that actively participate in intelligence itself.

AI is therefore not only a technological revolution.

It is a civilizational transformation.

14.10.3 The Institutional Crisis of the AI Age

Despite AI's transformative power, existing institutions remain misaligned with the demands of intelligent systems.

Today's institutional landscape can be broadly categorized into three domains:

Technological Institutions
Focused on innovation speed and engineering excellence.

Limitation:
Technology advances rapidly without sufficient ethical alignment or societal coordination.

Political Institutions
Focused on regulation, national interests, and policy enforcement.

Limitation:
Governance reacts slowly to globally evolving intelligent technologies.

Academic Institutions
Focused on research and knowledge creation.

Limitation:
Research often remains disconnected from real-world deployment, industry adoption, and governance implementation.

No single institution currently integrates technology, ethics, sustainability, and societal governance simultaneously.

The AI era therefore faces an institutional vacuum.

14.10.4 Why AI Requires a Civilizational Institution

Artificial Intelligence demands a new institutional model capable of integrating multiple dimensions at once:

- technological innovation,
- ethical responsibility,
- sustainability principles,
- governance coordination,
- human development.

Such an institution must transcend traditional boundaries.

It cannot be:

- merely technological,
- purely political,
- or exclusively academic.
- It must be civilizational.

What Is a Civilizational Institution?
A civilizational institution aligns intelligence with humanity's long-term survival and flourishing.

It ensures innovation continually answers fundamental questions:

- How should intelligence serve society?
- What responsibilities accompany technological power?
- How do we protect future generations?
- How do we sustain planetary balance?

Civilizational institutions guide direction, not merely operations.

14.10.5 The Emergence of Sustainable AI Institutions

Sustainable AI introduces responsibilities never before required of human institutions:

- safeguarding human dignity,
- protecting environmental ecosystems,
- maintaining democratic trust,
- ensuring equitable access to intelligence.

Without such institutions, AI risks becoming:

- economically powerful,
- socially destabilizing,
- environmentally unsustainable.

Sustainable AI therefore requires institutions explicitly designed for Responsible Intelligence Governance.

Example - Global AI Deployment

A powerful AI system is released worldwide.

Without a civilizational institution:

- companies compete independently,

- governance standards diverge,
- societal risks multiply.

With a Sustainable AI institution:

- shared ethical frameworks emerge,
- sustainability metrics guide deployment,
- collaborative governance protects humanity.

Institutions transform intelligence into collective progress.

14.10.6 The ZHA Global Institute - A New Institutional Model

The ZHA Global Institute represents this emerging institutional archetype.

It integrates four traditionally disconnected domains:

ZHA does not replace existing institutions.

It aligns and connects them into a coherent Sustainable AI ecosystem.

ZHA's Institutional Purpose

The Institute exists to:

- guide Sustainable AI transformation,
- develop responsible innovators,
- align global governance frameworks,
- cultivate civilization stewardship.

It operates as a coordinating intelligence layer for humanity.

14.10.7 Institutions as Guardians of Intelligence

In the age of Artificial Intelligence, institutions become guardians of intelligence itself.

Their role is not to control innovation, but to ensure intelligence evolves responsibly.

ZHA performs three civilizational functions:

1. Alignment
Ensuring AI development aligns with human values and sustainability principles.

2. Integration
Connecting education, industry, governance, and society into a unified ecosystem.

3. Stewardship
Protecting long-term civilization stability.

Example - Sustainable AI Education Pipeline

Students trained under ZHA Sustainability Mindset programs enter industry with ethical awareness.

Organizations adopt responsible innovation practices. Governments observe improved societal outcomes.

Institutional alignment emerges organically.

This is civilization stewardship in action.

14.10.8 From Organizations to Civilization Infrastructure

Traditional institutions manage sectors.

Civilizational institutions guide humanity.

The ZHA Global Institute operates as:

- a governance catalyst,
- a sustainability knowledge hub,
- a leadership development platform,
- a global coordination framework.

It treats intelligence itself as shared human infrastructure.

14.10.9 The ZHA Insight - Intelligence Needs Guidance

Humanity's greatest challenge is no longer creating intelligence.

It is guiding intelligence wisely.

Technology without institutional wisdom accelerates risk.

Institutional wisdom transforms intelligence into sustainable progress.

The future of Artificial Intelligence will therefore be shaped not only by engineers or entrepreneurs, but by institutions capable of integrating intelligence with responsibility.

Chapter Reflection

Technological revolutions succeed only when supported by institutions capable of guiding them responsibly.

Artificial Intelligence now requires a new type of institution:

- not merely technological,
- not purely political,
- not exclusively academic,
- but civilizational.

The ZHA Global Institute represents this new institutional model - ensuring Sustainable AI with Human-Centric Purpose evolves from philosophy into global reality.

AI transformation guided only by technology creates disruption.
AI transformation guided by institutions creates civilization.

Through ZHA, humanity transitions toward a conscious, sustainable, and intelligent civilization.

CHAPTER XV

Chapter 15: ZHA Sustainable AI Civilization Movement

Introduces institutional structures designed to scale Sustainable AI globally through governance, research, and collaborative leadership a Human-Centric Intelligent Civilization.

15.1 Core Purpose of the ZHA Sustainable AI Civilization Movement

Scaling Sustainable AI with Human-Centric Purpose Across Humanity

The ZHA Sustainable AI Civilization Movement exists to address a defining reality of the 21st century:

Artificial Intelligence is no longer a technological choice. It is becoming the operating intelligence of civilization itself.

As AI integrates into education, governance, healthcare, economies, and daily life, humanity faces a critical responsibility:

Intelligence must evolve alongside wisdom.

The core purpose of the ZHA movement is therefore clear:

To scale Sustainable AI with Human-Centric Purpose globally ensuring that technological progress strengthens humanity, society, and the planet simultaneously.

From Technology Transformation to Civilization Transformation

Most AI initiatives today focus on:

- automation efficiency,
- productivity enhancement,

- digital modernization,
- competitive advantage.

ZHA reframes AI transformation fundamentally.

AI transformation is not merely organizational change.

It is civilizational evolution.

Under ZHA philosophy:

- technology becomes a means,
- humanity becomes the purpose,
- sustainability becomes the direction.

AI transformation becomes:

Human Evolution Through Responsible Intelligence

Objective 1 - Embedding Sustainability Mindset Across Education

Civilizations are shaped first in classrooms before they are shaped in boardrooms.

The long-term success of Sustainable AI depends not on regulation alone, but on the mindset of future generations.

The ZHA movement embeds sustainability mindset within education systems worldwide.

Educational Transformation Includes:

- teaching ethical technology awareness,
- cultivating environmental responsibility,
- promoting collaboration over competition,
- nurturing long-term thinking in decision-making.

Students learn not only how to build technology, but how technology should serve humanity.

Example - ZSPC Educational Model

Students participating in sustainability certification programs:

- design community solutions,
- understand environmental impact,
- learn AI responsibility principles,
- practice collaborative leadership.

Outcome:

Graduates enter society as Sustainability Mindset Practitioners, capable of guiding intelligent systems responsibly.

Education becomes civilization preparation.

Objective 2 - Enabling Responsible Innovation Practices

Innovation without responsibility creates disruption.

Responsible innovation ensures technological advancement strengthens society rather than destabilizes it.

The ZHA movement enables organizations to adopt innovation models grounded in:

- ethical foresight,
- sustainability impact assessment,
- human-centered design,
- long-term societal value.

Example - Responsible AI Product Development

Traditional innovation question:
Can we build this?

ZHA innovation question:
Should we build this, and how does it benefit humanity sustainably?

Organizations guided by ZHA principles evaluate:

- environmental cost of AI models,
- social impact of automation,
- fairness of algorithms,
- transparency of decision systems.

Innovation evolves from invention toward stewardship.

Objective 3 - Aligning AI Development with ESG Principles

Environmental, Social, and Governance (ESG) frameworks traditionally function as reporting mechanisms.

ZHA transforms ESG into an operational intelligence system embedded directly into AI transformation.

Environmental Alignment

AI development prioritizes:

- energy efficiency,
- resource optimization,
- carbon-aware infrastructure.

Social Alignment

AI systems promote:

- inclusion,
- accessibility,
- equitable opportunity distribution.

Governance Alignment

AI deployment ensures:

- transparency,
- accountability,
- ethical oversight.

ESG becomes proactive behavior rather than reactive compliance.

Example - Sustainable AI Deployment

A company adopting ZHA alignment:

- optimizes models for energy efficiency,
- reskills employees impacted by automation,
- publishes transparent algorithm policies.

Outcome: Innovation increases while ESG risk decreases.

Sustainability becomes competitive advantage.

Objective 4 - Guiding Institutions Toward Civilization Stewardship

The AI era requires institutions to evolve from regulators and operators into stewards of civilization.

ZHA guides institutions - including governments, universities, corporations, and communities toward long-term responsibility.

Civilization stewardship means:

- thinking beyond quarterly performance,
- protecting future generations,
- aligning innovation with planetary limits,
- ensuring intelligence benefits all humanity.

Institutional Transformation

Institutions become guardians of intelligent civilization.

Sustainable AI with Human-Centric Purpose

At the heart of the ZHA movement lies a unifying philosophy: Sustainable AI with Human-Centric Purpose

This philosophy ensures that Artificial Intelligence:

- enhances human creativity rather than replacing it,
- strengthens communities rather than dividing them,
- protects ecosystems rather than exploiting them,
- supports democracy rather than undermining trust.

Humanity remains at the center of intelligence evolution.

From Intelligence Creation to Responsible Intelligence Evolution

Human history has progressed through expanding intelligence:

- survival intelligence,
- industrial intelligence,
- digital intelligence.

The ZHA movement introduces the next stage:
Responsible Intelligence.

AI transformation guided by ZHA enables humanity to evolve consciously rather than reactively. Technology becomes an instrument of human maturity.

The ZHA Insight

The future of civilization will not be determined by how intelligent machines become.

It will be determined by how responsibly humanity chooses to use intelligence.

The ZHA Sustainable AI Civilization Movement exists to ensure that:

- innovation serves life,
- intelligence protects humanity,
- progress sustains the planet.

AI transformation therefore becomes more than technological advancement.

It becomes human evolution guided by responsibility.

15.5 Institutional Architecture of the ZHA Sustainable AI Civilization Movement

Designing the Operating Structure of a Human-Centric Intelligent Civilization

Civilization movements succeed only when supported by strong institutional architecture.

Ideas inspire change.
Institutions sustain change.

The ZHA Sustainable AI Civilization Movement therefore operates through an interconnected ecosystem of institutions designed to scale Sustainable AI with Human-Centric Purpose globally. Unlike traditional hierarchical organizations, the ZHA architecture functions as a living ecosystem, where education, governance, research, industry, and society evolve together. Each institutional component performs a distinct civilizational role while remaining deeply interconnected. Together, they form the operational foundation of Sustainable AI civilization.

1. ZHA Global Institute - The Intellectual and Governance Nucleus

The ZHA Global Institute serves as the central coordinating intelligence of the movement.

It is not merely a research institute or policy body.

It functions as the civilization design center for Sustainable AI transformation.

Core Responsibilities:

- Develop Sustainable AI frameworks and methodologies
- Guide global AI governance principles
- Conduct interdisciplinary sustainability research
- Train future civilization leaders
- Align institutions across countries and sectors

The Institute integrates knowledge from:

- technology,
- ethics,
- sustainability science,
- social systems,
- governance studies.

Role in AI Transformation

Traditional institutions study technology after it emerges.

The ZHA Global Institute anticipates technological evolution before disruption occurs.

It enables:

- proactive governance,
- sustainability-aligned innovation,
- long-term civilization foresight.

Example

When new generative AI capabilities emerge, the Institute:

- evaluates environmental impact,
- assesses social consequences,
- recommends governance frameworks,
- guides organizations toward responsible deployment.

The Institute becomes humanity's strategic compass for intelligence evolution.

2. ZHA World Council - Civilization-Scale Leadership & Foresight Body

The ZHA World Council represents the leadership dimension of the movement.

While the Institute generates knowledge, the World Council guides civilization direction.

It brings together:

- global leaders,
- policymakers,
- scientists,
- educators,
- sustainability practitioners,

- ethical thinkers.

Core Functions

- Establish global Sustainable AI vision
- Facilitate international collaboration
- Provide civilization foresight planning
- Resolve cross-sector ethical dilemmas
- Promote global alignment on AI stewardship

The Council operates beyond national interest toward planetary responsibility.

Civilization Foresight

The Council evaluates long-term questions such as:

- How will AI reshape employment?
- How should intelligence be governed globally?
- How can AI protect planetary ecosystems?
- How do we ensure inclusive technological futures?

Leadership evolves from authority to stewardship.

Example

Before large-scale autonomous systems are deployed globally, the Council develops shared ethical guidelines ensuring:

- public safety,
- employment transition support,
- social trust preservation.

Civilization decisions become collaborative rather than competitive.

3. ZHA Sustainable AI Research Network - Global Intelligence Collaboration

Innovation in the AI era cannot remain isolated.

The ZHA Sustainable AI Research Network connects global knowledge ecosystems into a collaborative intelligence system.

Participants include:

- universities,
- research laboratories,
- innovation hubs,
- sustainability organizations,
- technology communities.

Research Focus Areas:

- Sustainable AI architectures
- Low-energy machine learning models
- Ethical algorithm design
- Human-AI collaboration systems
- Climate intelligence platforms
- Social impact analytics

Research becomes civilization infrastructure rather than academic output alone.

Example - Collaborative Climate AI Research

Researchers across continents collaborate to build shared AI models predicting climate risks.

Results include:

- reduced duplicated research effort,
- faster global climate response,
- open sustainability innovation.

Knowledge becomes a shared human resource.

4. ZHA Education & Certification Ecosystem (ZSPC)

Developing Sustainability Practitioners Across Generations

Civilization transformation begins with education. The ZHA Sustainability Practitioners Community (ZSPC) serves as the human development engine of the movement. Its purpose is to cultivate individuals capable of guiding intelligent systems responsibly.

Participants:

- school students,
- university learners,
- educators,
- professionals,
- institutional leaders.

Educational Philosophy

ZSPC moves education beyond knowledge acquisition toward mindset formation.

Learners develop:

- sustainability thinking,
- ethical technology awareness,
- collaboration skills,
- community responsibility,
- long-term decision intelligence.

Outcome:
A globally aligned workforce of Sustainability Mindset Practitioners.

Instead of adapting to technology blindly, society develops citizens capable of shaping technology consciously.

Example:

Students conduct sustainability projects integrating AI solutions for local community challenges:

- water conservation analytics,
- waste reduction systems,
- energy optimization solutions.

Education becomes civilization participation.

5. ZHA Industry & Innovation Consortium - Industry as Civilization Partner

Industry plays a decisive role in shaping technological futures. The ZHA Industry & Innovation Consortium aligns corporations toward Sustainable AI transformation. Rather than treating industry as a source of disruption, ZHA positions industry as a partner in civilization progress.

Consortium Objectives

- embed ESG principles into product development,
- standardize responsible AI practices,
- promote collaborative innovation,
- reduce environmental impact of technology,
- encourage ethical digital economies.

Transformation of Corporate Purpose

Example - Sustainable AI Product Ecosystem

Technology companies collaborate through ZHA frameworks to:

- optimize energy consumption of AI models,
- share ethical design standards,
- co-develop sustainability metrics.

Industry transitions from economic driver to civilization steward.

Interconnected Institutional Ecosystem

The strength of the ZHA architecture lies in integration.

Each structure reinforces the others:

- Institute provides knowledge and governance design.
- World Council provides leadership and foresight.
- Research Network generates collaborative intelligence.
- ZSPC Ecosystem develops responsible human capability.
- Industry Consortium operationalizes Sustainable AI globally.

Together, they create a self-reinforcing civilization system.

From Institutions to Civilization Infrastructure

Traditional institutions operate independently.

The ZHA architecture operates as an integrated civilization operating system.

- Education informs innovation.
- Innovation aligns with governance.
- Governance protects society.
- Society sustains civilization.

Sustainable AI becomes embedded across human systems.

The ZHA Insight

Civilizations do not transform through technology alone. They transform when institutions align around shared purpose.

The ZHA Sustainable AI Civilization Movement builds the institutional architecture required for humanity's next evolutionary stage:
A Human-Centric Intelligent Civilization guided by Sustainable AI.

15.6 Sustainable AI Governance Model - Distributed Civilization Stewardship

From Centralized Control to Collective Responsibility

Artificial Intelligence challenges traditional models of governance.

Historically, governance operated through centralized authority:

- governments regulated industries,
- institutions enforced rules,
- organizations complied with standards.

This model functioned effectively during slower technological eras.

However, Artificial Intelligence evolves faster than centralized systems can respond.

AI systems:

- learn continuously,
- operate globally,
- influence society instantly,
- cross institutional and national boundaries.

No single authority can fully govern intelligence at planetary scale.

The AI era therefore demands a new governance philosophy:
Distributed Civilization Stewardship

The ZHA Sustainable AI Governance Model introduces governance as a shared responsibility across humanity.

Why Traditional Governance Is No Longer Sufficient

Traditional governance follows a reactive pattern:

- Technology emerges.
- Harm appears.
- Regulation follows.

This approach creates continuous lag between innovation and protection.

Examples include:

- privacy laws introduced after data misuse,
- social media governance after societal disruption,
- AI ethics guidelines emerging after algorithmic bias incidents.

Reactive governance cannot stabilize exponentially evolving intelligence systems.

Sustainable AI requires anticipatory governance.

The ZHA Governance Paradigm

ZHA redefines governance from authority-based regulation to stewardship-based collaboration.

Governance becomes:

- proactive rather than reactive,
- participatory rather than centralized,
- value-driven rather than compliance-driven,
- civilization-oriented rather than institution-oriented.

The core principle:
Intelligence must be guided collectively by those who create, use, and are affected by it.

Distributed Governance Structure

The ZHA model distributes responsibility across four collaborating pillars.

1. Institutions Collaborate
Academic institutions, research bodies, and knowledge organizations provide intellectual leadership.

They:

- develop ethical frameworks,
- conduct sustainability research,
- assess long-term societal impact,
- guide evidence-based decision-making.

Institutions become knowledge stewards.

Example:
Universities collaborate globally through ZHA networks to evaluate environmental impact of large AI models before widespread deployment.

Governance begins before disruption occurs.

2. Communities Participate
Communities are no longer passive recipients of technological change.

Under ZHA governance, citizens actively shape AI adoption.

Community participation includes:

- public consultation,
- participatory technology design,
- local sustainability initiatives,
- social impact feedback systems.

Technology becomes socially accountable.

Example:
Before implementing AI-driven urban surveillance systems, communities participate in discussions on privacy, safety, and fairness.

Trust replaces resistance.

3. Industries Contribute
Industry holds the greatest operational influence over AI deployment.

ZHA governance positions corporations not merely as innovators but as civilization partners.

Industry responsibilities include:

- embedding ESG principles into SDLC,
- transparent algorithm design,
- sustainable infrastructure adoption,
- workforce transition support.

Corporate success aligns with societal wellbeing.

Example:
An AI company deploying automation reskills employees into AI supervision roles instead of workforce displacement.

Innovation strengthens society rather than destabilizing it.

4. Governments Align
Governments remain essential but evolve from regulators to coordinators.

Their role becomes:

- aligning international standards,
- enabling collaborative governance,
- supporting Sustainable AI policies,
- ensuring equitable access to intelligence.

Governments guide ecosystems rather than controlling innovation alone.

Example:
Multiple nations adopt shared Sustainable AI frameworks aligned with ZHA principles, enabling cross-border ethical AI deployment.

Governance becomes cooperative rather than competitive.

Governance Layers of the ZHA Model

Distributed stewardship operates across three interconnected layers.

Layer 1 - Individual Governance: Sustainability Mindset

The foundation of Sustainable AI governance begins with individuals.

Policies cannot replace human responsibility.

When individuals adopt sustainability mindset:

- ethical thinking precedes regulation,
- responsible decisions occur naturally,
- innovation aligns with long-term wellbeing.
- Individuals become self-governing innovators.

Example:

A developer chooses energy-efficient algorithms voluntarily, recognizing environmental responsibility.

Governance exists before rules are required.

Layer 2 - Organizational Governance: Responsible SDLC & AI ESG

Organizations operationalize responsibility through structured systems.

Key mechanisms include:

- Responsible Software Development Life Cycle (SDLC),
- ESG-integrated product development,
- ethical AI audits,
- sustainability performance metrics.

Governance becomes embedded into operational workflows.

Example:
An organization integrates ESG checkpoints into every development sprint:

- environmental impact review,
- fairness validation,
- transparency documentation.

Governance becomes part of innovation itself.

Layer 3 - Global Governance: Coordinated Sustainable AI Policies

At the global level, governance aligns institutions and nations around shared principles.

Objectives include:

- interoperable ethical standards,
- coordinated AI safety practices,

- global sustainability metrics,
- collaborative research governance.

Civilization moves toward shared intelligence stewardship.

Example:
Global Sustainable AI standards ensure healthcare AI systems maintain consistent fairness and safety regardless of country.

Human dignity becomes universal.

Governance Becomes Collective Stewardship

The ZHA insight transforms governance fundamentally:

- Governance is no longer about control.
- It becomes stewardship.

Stewardship means:

- protecting future generations,
- balancing innovation with responsibility,
- sustaining planetary systems,
- guiding intelligence toward human flourishing.

Every participant becomes a steward:

- individuals,
- organizations,
- institutions,
- nations.

From Regulation to Civilization Stewardship

The result is a resilient governance ecosystem capable of guiding intelligent civilization.

The ZHA Insight

Artificial Intelligence cannot be governed by rules alone.

It must be guided by shared responsibility.

Distributed Civilization Stewardship ensures that:

- intelligence evolves ethically,
- innovation remains sustainable,
- society retains agency,
- civilization remains stable.

The future of governance is not stronger control.

It is wiser collaboration.

15.7 Research as Sustainable AI Civilization Intelligence

Transforming Knowledge into Stewardship for a Human-Centric Intelligent Civilization

Human civilization has always progressed through research.

Scientific discovery enabled agriculture.
Engineering research enabled industry.
Digital research enabled the information age.

Yet historically, research followed a limited purpose:

- generate knowledge,
- publish findings,
- advance academic reputation,
- support technological innovation.

In the age of Artificial Intelligence, this model becomes insufficient. AI systems influence civilization itself. Therefore, research must evolve. Research is no longer only about understanding the world. It becomes about guiding humanity responsibly within an intelligent world.

From Academic Research to Civilization Intelligence

Traditional research measures success through:

- publications,
- patents,
- technological breakthroughs,
- institutional recognition.

The ZHA Sustainable AI Civilization Movement introduces a new paradigm:

Research as Civilization Intelligence Creation

Civilization intelligence means knowledge developed explicitly to:

- protect humanity,
- sustain ecosystems,
- strengthen societies,
- guide intelligent technologies responsibly.

Research becomes an act of stewardship rather than competition.

Why Research Must Transform in the AI Era

Artificial Intelligence introduces unprecedented characteristics:

- systems evolve autonomously,
- algorithms influence human behavior,
- decisions scale globally,
- impacts unfold across generations.

Research conducted without sustainability awareness risks creating unintended consequences.

Examples include:

- energy-intensive AI models increasing carbon emissions,
- biased datasets reinforcing inequality,
- automation research ignoring workforce transition.

The challenge is no longer discovering capability.

The challenge is ensuring capability aligns with civilization wellbeing.

ZHA Research Philosophy
The ZHA model redefines research through a guiding principle:

Knowledge must serve life, society, and planetary sustainability.

Researchers become:

- foresight thinkers,
- ethical designers,
- sustainability stewards,
- civilization architects.

Knowledge becomes responsibility.

Core Focus Areas of Sustainable AI Civilization Intelligence

1. Sustainable AI Architectures
AI systems require immense computational power.

Without sustainable design, intelligence growth increases environmental burden.

ZHA research promotes:

- energy-efficient model design,
- low-carbon computing infrastructure,
- adaptive resource optimization,
- decentralized intelligent systems.

Example:
Researchers develop smaller, optimized AI models delivering equivalent performance using significantly less energy.

Outcome:
Innovation continues while environmental impact decreases.
AI becomes environmentally conscious intelligence.

2. Ethical Algorithm Design
Algorithms increasingly influence:

- hiring decisions,
- healthcare diagnostics,
- financial access,
- public information.

ZHA research embeds ethics directly into algorithm creation.

Focus areas include:

- fairness validation,
- bias mitigation,
- transparency mechanisms,
- explainable AI systems.

Example:
A healthcare AI model undergoes fairness testing across demographic groups before deployment, ensuring equitable treatment outcomes.

Ethics shifts from post-deployment correction to design principle.

3. Societal Impact Modeling
AI decisions ripple across entire societies.

ZHA research expands beyond technical performance to evaluate:

- employment transformation,
- psychological impact,
- social cohesion,
- governance implications.

Researchers simulate long-term societal outcomes before technology deployment.

Example:
Before introducing large-scale automation in logistics, researchers model workforce transitions and propose reskilling ecosystems.

Technology deployment becomes socially responsible planning.

4. Climate Intelligence
The climate crisis represents humanity's greatest collective challenge.

AI provides unprecedented capability to understand and respond to planetary systems.

ZHA research advances:

- environmental prediction models,
- biodiversity monitoring systems,
- climate risk intelligence platforms,
- sustainable agriculture analytics.

AI becomes an instrument of planetary stewardship.

Example:
AI systems analyze satellite data to predict deforestation risks and enable early intervention policies.
Research protects ecosystems rather than exploiting them.

5. Human Wellbeing Analytics
Technological progress must ultimately improve human life.

ZHA research expands measurement beyond economic indicators toward human wellbeing.

Areas include:

- mental health analytics,
- education accessibility intelligence,
- community resilience modeling,
- quality-of-life indicators.

Human flourishing becomes a core research objective.

Example:
AI-driven public health research identifies social isolation patterns and guides community interventions improving societal wellbeing.

Technology serves humanity directly.

Knowledge Becomes Stewardship

The ZHA Civilization Movement reframes the role of researchers.

Researchers are no longer only:

- scientists,
- engineers,
- analysts.

They become guardians of future civilization.

Every research decision asks:

- Does this strengthen humanity?
- Does this sustain the planet?
- Does this promote fairness?
- Does this protect future generations?

Knowledge becomes stewardship.

Collaborative Research Ecosystem

Sustainable AI Civilization Intelligence cannot emerge from isolated institutions.

ZHA promotes collaborative research networks connecting:

- universities,
- industry laboratories,
- policymakers,
- sustainability practitioners,
- communities.

Research becomes a shared global endeavor.

Example: Global Sustainable AI Collaboration

Multiple universities and technology organizations collaborate to create open standards for low-energy AI training methods.

Benefits include:

- reduced environmental impact,
- accelerated innovation,

- shared civilizational progress.

Competition gives way to collaboration.

Research as Civilization Infrastructure

Under ZHA philosophy, research becomes infrastructure similar to:

- transportation systems,
- healthcare networks,
- energy grids.

It continuously supports civilization stability.

Research guides:

- ethical innovation,
- sustainable economic systems,
- responsible governance,
- societal resilience.

Civilization learns before crisis emerges.

The ZHA Insight

The future of humanity depends not only on intelligent machines, but on intelligent research guiding them.

Civilizations decline when knowledge advances without wisdom.

Civilizations flourish when knowledge serves collective wellbeing.

Sustainable AI Civilization Intelligence ensures that research becomes humanity's most powerful tool for conscious evolution.

15.8 Collaborative Leadership for the AI Age

From Management Authority to Civilization Stewardship

Every era of civilization produces the leaders it requires.

- The agricultural age required community organizers.
- The industrial age required operational managers.
- The digital age required innovation executives.

The Artificial Intelligence era demands something fundamentally different:

Leaders capable of guiding intelligence itself.

AI transforms not only technology and economies, but decision-making across civilization.
Leadership can no longer focus solely on efficiency, growth, or control.

The AI era requires leaders who understand interconnected systems, long-term consequences, and shared human responsibility.

Leadership evolves from management toward stewardship.

Why Traditional Leadership Models Are Insufficient

Traditional leadership models were built for predictable environments.

They emphasized:

- hierarchy,
- authority,
- competition,
- short-term performance optimization.

These models succeeded when change occurred slowly.

Artificial Intelligence introduces:

- exponential innovation,
- global interdependence,
- ethical complexity,
- societal-scale impact.

Decisions made by leaders today influence millions instantly.

Leadership without expanded awareness risks accelerating disruption.

The AI era therefore requires collaborative leadership.

The ZHA Leadership Paradigm

The ZHA Sustainable AI Civilization Movement introduces a new leadership archetype grounded in four core capabilities:

- Systems Thinking
- Sustainability Advocacy
- Ethical Innovation

Civilization Stewardship
Together, these qualities transform leaders into guardians of intelligent civilization.

1. Systems Thinkers

AI operates within interconnected systems:

- technological systems,
- economic systems,
- ecological systems,
- social systems.

A decision in one domain produces ripple effects across others.

Leaders must therefore think beyond isolated outcomes.

They must understand relationships, feedback loops, and long-term systemic consequences.

Example: AI Automation Decision

A traditional leader asks:
How much cost reduction will automation achieve?

A systems-thinking leader asks:

- How will automation affect workforce stability?
- What reskilling pathways are required?
- What social impact emerges long term?
- How does this decision affect community resilience?

The decision becomes holistic rather than operational.

Systems thinking transforms leadership from problem-solving into future-shaping.

2. Sustainability Advocates

Sustainability is no longer an environmental initiative.

It becomes a leadership responsibility.

AI systems consume energy, reshape economies, and influence human behavior.

Leaders must advocate for sustainability across all dimensions:

- environmental sustainability,

- social inclusion,
- economic resilience,
- psychological wellbeing.

Example: Sustainable AI Deployment

Instead of deploying the largest possible AI model, a sustainability-focused leader prioritizes:

- energy-efficient architecture,
- optimized infrastructure,
- responsible data usage.

Innovation continues while planetary impact decreases.

Leadership aligns progress with preservation.

3. Ethical Innovators

Innovation without ethics creates unintended harm.

Ethical innovators integrate moral reflection into creation itself.

They ask not only:
Can we build this?

but also:
Should we build this, and who benefits or suffers?

Ethical innovation anticipates consequences before deployment.

Example: AI Healthcare System

An ethical leader ensures that medical AI systems:

- undergo fairness testing,
- remain explainable to doctors and patients,

- protect privacy,
- maintain human oversight.

Technology enhances trust rather than replacing human judgment.

Ethics becomes an innovation accelerator rather than a restriction.

4. Civilization Stewards
The highest evolution of leadership in the AI era is stewardship.

Stewards recognize that leadership decisions affect:

- future generations,
- planetary ecosystems,
- societal stability,
- cultural evolution.

Leadership moves beyond ownership toward guardianship.

Civilization Stewardship Mindset

A civilization steward asks:

- Does this innovation strengthen humanity?
- Does it preserve environmental balance?
- Will future generations benefit from this decision?
- Does it increase collective wellbeing?

Leadership expands from organizational success to civilizational responsibility.

Leadership Evolution - From Management to Stewardship

Management optimizes systems.
Stewardship protects civilization.

Collaborative Leadership - The ZHA Approach

The complexity of AI governance exceeds individual leadership capacity.

No single leader or organization can guide intelligent civilization alone.

ZHA therefore promotes collaborative leadership ecosystems.

Leadership becomes distributed across:

- educators,
- scientists,
- entrepreneurs,
- policymakers,
- community leaders,
- sustainability practitioners.

Decision-making becomes collective intelligence.

Example - ZHA Leadership Fellowship

Leaders participating in ZHA programs learn to:

- integrate ESG into AI strategies,
- collaborate across sectors,
- balance innovation with societal stability,
- design long-term civilization outcomes.

Leaders transition from executives to civilization stewards.

Emotional Intelligence and Conscious Leadership

AI amplifies analytical intelligence.

Human leaders must amplify emotional and ethical intelligence.

Future leaders must cultivate:

- empathy,
- humility,
- adaptability,
- reflective thinking.

Human consciousness becomes the balancing force to artificial intelligence.

Leadership as a Cultural Force

Leadership shapes organizational culture, and culture shapes civilization.

When leaders embody sustainability mindset:

- employees adopt responsibility,
- organizations innovate ethically,
- societies trust technology.

Culture becomes the invisible governance system of Sustainable AI civilization.

The ZHA Insight

The future of Artificial Intelligence will not be determined solely by algorithms.

It will be determined by the quality of leadership guiding those algorithms.

Humanity does not need more powerful leaders.

It needs wiser leaders.

Leaders who collaborate rather than dominate.
Leaders who steward rather than control.

Leaders who guide intelligence toward human flourishing.

15.9 Scaling Sustainable AI Globally

How Civilization Transformation Spreads Through Network Intelligence

Civilizations do not change through orders or enforcement.

They evolve through adoption.

Throughout history, transformative ideas spread not because they were mandated, but because they became meaningful, useful, and inevitable.

Democracy spread through shared aspiration.
Education spread through societal necessity.
Digital technology spread through network effects.

Sustainable AI civilization follows the same pattern.

The ZHA Sustainable AI Civilization Movement scales globally through network intelligence interconnected adoption across education, professions, organizations, and governance systems.

Transformation becomes organic rather than imposed.

The Principle of Network Effects in Civilization Change

A network effect occurs when each new participant strengthens the entire system.

In technological platforms:
more users create more value.

In civilization movements:
more conscious participants create systemic transformation.

ZHA applies this principle to Sustainable AI adoption.

Every new participant becomes both beneficiary and catalyst of change.

Four Pathways of Global Scaling

The ZHA movement expands through four mutually reinforcing pathways.

1. Schools Adopt Sustainability Mindset
Civilization transformation begins with education.

When schools integrate sustainability mindset:

- students learn responsible innovation,
- ethical awareness develops early,
- future professionals enter society prepared for Sustainable AI.

Education becomes the seed of long-term civilization stability.

Example - Sustainability Mindset Education

Students participating in ZSPC programs:

- analyze environmental challenges,
- develop community-based solutions,
- understand AI's societal impact,
- practice collaborative leadership.

Within a decade, these learners become:

- engineers,
- entrepreneurs,
- policymakers,
- researchers,
- already aligned with Sustainable AI values.

Transformation begins before disruption occurs.

2. Professionals Practice Responsible Innovation
Professionals translate mindset into real-world action.

As sustainability-trained individuals enter workplaces:

- AI projects integrate ethical considerations,
- development teams embed ESG principles,
- innovation decisions prioritize long-term impact.

Responsible innovation spreads horizontally across industries.

Example: Workplace Transformation

A software architect trained in sustainability mindset:

- selects energy-efficient cloud infrastructure,
- ensures fairness testing in algorithms,
- promotes transparent decision systems.

One professional influences an entire organization.

Thousands influence industries.

3. Organizations Deploy Sustainable AI
Organizations operationalize civilization change.

When companies adopt Sustainable AI:

- ESG becomes embedded in SDLC,
- innovation aligns with social wellbeing,
- technology strengthens public trust.

Corporate transformation creates economic momentum for Sustainable AI.

Example: Sustainable Enterprise Adoption

An enterprise adopting ZHA framework:

- integrates ESG checkpoints into product lifecycle,
- reskills employees impacted by automation,
- reduces carbon footprint of AI infrastructure.

Business success and sustainability become aligned rather than competing goals.

4. Governments Align Governance Frameworks
Governments play a coordinating role rather than a controlling one.

As education, professionals, and organizations evolve, governments naturally align policies to support emerging practices.

Policy follows culture.

Governments enable:

- Sustainable AI standards,
- cross-border collaboration,
- ethical innovation incentives,
- inclusive technology access.

Governance becomes facilitator of civilization evolution.

Example: National AI Alignment

A country adopting ZHA-aligned policies:

- introduces sustainability mindset education nationally,
- supports ethical AI startups,
- aligns ESG metrics with innovation funding.

Technology strengthens national resilience.

Organic Change Rather Than Enforcement

Traditional transformation relies on regulation and enforcement.

The ZHA Civilization Movement operates differently.

Change spreads through:

- inspiration rather than compulsion,
- participation rather than compliance,
- alignment rather than control.

People adopt Sustainable AI because it works - ethically, economically, and socially.

Civilization evolves voluntarily.

Network Intelligence - The Scaling Mechanism

Each participant strengthens the ecosystem:

- schools create responsible citizens,
- professionals transform workplaces,
- organizations reshape economies,
- governments stabilize systems.

The movement grows exponentially.

Transformation becomes self-sustaining.

Global Ripple Effect of Sustainable AI

Small local adoption creates global impact.

A single school adopting sustainability mindset influences future leaders.
A single company adopting Sustainable AI influences industry standards.
A single government aligning policies influences international cooperation.

Networked change scales civilization.

Sustainable AI as a Shared Global Language

As adoption grows, Sustainable AI becomes a common framework understood across cultures and nations.

Organizations begin speaking the same language of:

- responsibility,
- sustainability,
- human-centered innovation,
- civilization stewardship.

This shared understanding reduces global technological conflict.

From Innovation Diffusion to Civilization Evolution

The ZHA insight reveals a profound truth:
Civilization transformation does not occur when everyone changes at once.

It occurs when enough connected nodes change to influence the whole system.

Scaling Sustainable AI globally means enabling millions of aligned decisions rather than enforcing one centralized rule.

The ZHA Insight

Civilizations do not scale through authority.

They scale through alignment.

When education, industry, governance, and society move together, transformation becomes inevitable.

The ZHA Sustainable AI Civilization Movement spreads because it aligns with humanity's deepest aspiration: to progress without destroying the future.

15.10 The ZHA Civilization Index

Measuring Progress in the Age of Sustainable AI Civilization

For centuries, human progress has been measured primarily through economic indicators.

Gross Domestic Product (GDP) became the dominant metric for evaluating national success.

Economic growth was assumed to represent societal advancement.

However, the Artificial Intelligence era exposes the limitations of purely economic measurement.

A society may achieve:

- high productivity,
- advanced technology,
- economic expansion,

yet still experience:

- environmental collapse,
- social inequality,
- declining trust,
- psychological distress,
- governance instability.

Economic growth alone no longer reflects civilization wellbeing.

The AI age demands a new measurement system.

From Economic Output to Civilization Health

Artificial Intelligence amplifies whatever societies choose to measure.

If success is measured only by productivity, AI accelerates exploitation.
If success is measured by wellbeing, AI strengthens civilization.

The ZHA Sustainable AI Civilization Movement therefore introduces a new framework:

The ZHA Civilization Index

The Index shifts evaluation from economic output to civilization health.

It measures whether intelligence advances humanity sustainably across generations.

Why Civilization Needs a New Index

Traditional progress metrics were designed for earlier eras:

Artificial Intelligence affects:

- ecosystems,
- social structures,
- governance systems,
- human psychology.

Therefore, progress must be assessed holistically.

The ZHA Civilization Index evaluates civilization as a living system rather than an economic machine.

Core Dimensions of the ZHA Civilization Index

The Index measures five integrated dimensions.

1. Environmental Sustainability
A civilization cannot thrive if its ecological foundation collapses.

This dimension evaluates whether AI and technological innovation protect planetary systems.

Key Indicators:

- carbon footprint of AI infrastructure,
- energy efficiency of digital ecosystems,
- biodiversity preservation,
- resource conservation performance,
- climate resilience capability.

Example:
Two countries deploy advanced AI systems.

Country A maximizes computing scale without environmental consideration.

Country B optimizes AI models for energy efficiency.

Under the ZHA Civilization Index, Country B demonstrates higher civilization health despite similar technological capability.

Technology becomes environmentally responsible intelligence.

2. Social Inclusion
Sustainable civilization ensures technology benefits all people, not only privileged groups.

AI must reduce inequality rather than amplify it.

Key Indicators:

- accessibility of AI technologies,
- digital inclusion across socioeconomic groups,
- equitable education access,
- workforce transition support,
- gender and community representation.

Example:
Automation replaces certain jobs.

A civilization aligned with ZHA principles invests in reskilling programs, enabling displaced workers to transition into new roles.

Progress becomes inclusive rather than disruptive.

3. Governance Transparency
Trust becomes the most valuable currency of intelligent societies.

AI systems influencing decisions must remain transparent and accountable.

Key Indicators:

- algorithm transparency standards,
- ethical audit compliance,
- public accountability mechanisms,
- participatory governance engagement,
- corruption reduction through intelligent systems.

Example:
Public-sector AI decision systems provide explainable outcomes accessible to citizens.

Transparency strengthens democracy rather than weakening it.

Governance evolves into intelligent stewardship.

4. Ethical AI Adoption

Technological advancement must align with ethical responsibility.

This dimension evaluates whether AI deployment respects human dignity and societal stability.

Key Indicators

- fairness and bias mitigation,
- human oversight in critical decisions,
- responsible automation practices,
- ethical innovation frameworks,
- AI safety maturity levels.

Example:
Healthcare AI supports doctors rather than replacing them, ensuring human judgment remains central.

Technology augments humanity instead of dominating it.

5. Human Wellbeing Indicators

The ultimate purpose of civilization is human flourishing.

Economic success without wellbeing represents civilizational imbalance.

Key Indicators

- mental health outcomes,
- life satisfaction levels,
- education quality,
- community cohesion,
- sense of purpose and belonging.

Example:
A technologically advanced society experiencing widespread burnout scores lower on the Civilization Index than a balanced society prioritizing wellbeing.

Progress becomes human-centered rather than machine-centered.

Civilization Wellbeing - The New Definition of Success

The ZHA Civilization Index reframes success.

Traditional question:

- How fast are we growing?

ZHA question:

- Are we evolving sustainably as a civilization?

Success moves beyond GDP toward:

Civilization Wellbeing

A healthy civilization demonstrates:

- technological intelligence,
- environmental balance,
- social harmony,
- transparent governance,
- human fulfillment.

The Role of AI in Measuring Civilization Health

Artificial Intelligence itself becomes a tool for evaluating civilization progress.

AI systems analyze:

- environmental trends,
- social inclusion patterns,
- governance effectiveness,
- wellbeing indicators.

Intelligence measures intelligence responsibly.

Global Application of the ZHA Civilization Index

The Index can guide:

- national policy planning,
- corporate ESG strategies,
- educational transformation,
- global cooperation frameworks,
- Sustainable AI governance.

Countries, organizations, and institutions begin aligning progress around shared civilization goals.

From Competition to Collective Advancement

GDP encourages competition between nations.

The ZHA Civilization Index encourages collaboration.

Humanity recognizes that civilization health is interconnected:

- environmental damage crosses borders,
- technological risks affect all societies,
- human wellbeing is globally linked.

Progress becomes collective rather than competitive.

The ZHA Insight

Humanity has learned how to measure wealth.

The AI era requires learning how to measure wisdom.

The ZHA Civilization Index ensures that intelligence development strengthens civilization rather than destabilizing it.

What humanity measures determines what humanity becomes.

By measuring civilization wellbeing, humanity chooses sustainable evolution.

15.11 Sustainable AI as Civilization Infrastructure

From Technology Systems to Civilization Systems

Throughout history, civilizations have been shaped by infrastructure.

Infrastructure determines how societies function, how economies grow, and how humanity survives across generations.

Roads enabled trade.

Electricity enabled industry.

The internet enabled global communication.

Each technological era introduced a foundational layer upon which civilization evolved.

Artificial Intelligence now represents the next foundational layer.

But unlike previous infrastructures, AI does not merely move goods or information.

It shapes intelligence itself.

Under the ZHA Sustainable AI Civilization Movement, AI transitions from a technological capability into Civilization Infrastructure.

Technology no longer exists only to optimize efficiency.

Technology exists to ensure civilization continuity.

Why AI Must Become Sustainable Infrastructure

Today, AI influences nearly every human decision:

- education pathways,
- medical diagnoses,
- financial systems,
- employment markets,
- governance decisions,
- environmental management.

If AI infrastructure is built only for speed or profit, civilization risks instability.

If AI infrastructure is built sustainably, civilization gains resilience.

The ZHA insight is clear:

Intelligence must be governed as shared human infrastructure.

Sustainable AI ensures that intelligence strengthens humanity rather than destabilizes it.

The Four Pillars of Sustainable AI Civilization Infrastructure

CHAPTER XVI

Chapter 16: Civilization Charter and Constitution

Introduces institutional structures designed to scale Sustainable AI globally through governance, research, and collaborative leadership.

Human civilizations endure when values become institutions and principles become governance. Ideas inspire. Movements mobilize, But charters and constitutions sustain civilization across generations.

The ZHA Civilization Charter and Constitution represent humanity's transition from technological transformation toward conscious civilization governance.

They establish institutional structures capable of scaling Sustainable AI with Human-Centric Purpose globally through:

- governance,
- research,
- education,
- innovation,
- collaborative leadership.

This chapter introduces the framework through which humanity learns not only to create intelligence, but to govern intelligence responsibly.

16.1 Why Civilization Requires a Charter

Establishing Shared Purpose in the Age of Intelligent Systems

Every enduring civilization begins not with technology, wealth, or power, but with shared agreement.

Civilizations survive when people collectively decide:

- what they value,
- what they protect,
- what they permit,
- and what future they seek to build together.

These agreements become charters.

A charter represents civilization's declaration of intent, a conscious statement defining how humanity chooses to live, govern, and progress.

Technology creates possibility.

A charter defines responsibility.

The Role of Charters in Human Civilization

Throughout history, charters have appeared at moments when societies confronted transformational change.

When power expanded, humanity needed ethical boundaries.
When societies grew complex, humanity needed shared direction.
When new capabilities emerged, humanity needed collective responsibility.

Charters transformed uncertainty into stability.

They converted power into accountability.

Historical Foundations of Civilizational Charters

The Magna Carta - Limiting Power

In medieval England, authority concentrated heavily within monarchy.

The Magna Carta established a revolutionary idea:
power must operate within agreed principles.

It introduced:

- accountability,
- legal fairness,
- shared governance expectations.

The charter did not eliminate authority.

It civilized authority.

Constitutional Democracies - Protecting Human Rights

As nations modernized, constitutions emerged to define:

- individual freedoms,
- civic responsibility,
- governmental limits,
- rule of law.

These charters enabled societies to transition from rule by force toward governance by shared agreement.

Rights became institutionalized.

Trust became possible.

International Charters - Enabling Global Cooperation

In the modern era, globalization required agreements beyond national borders.

International charters enabled cooperation across humanity:

- human rights frameworks,
- environmental agreements,

- global peace institutions.

These agreements recognized an essential truth:

Some challenges exceed national boundaries.

Civilization must occasionally think collectively.

The AI Era - A New Civilizational Threshold

Humanity now stands at another historical inflection point.

Artificial Intelligence introduces capabilities unlike any previous technology.

AI systems:

- learn autonomously,
- generate knowledge,
- influence human behavior,
- guide decision-making,
- operate across the entire planet instantly.

For the first time, humanity is building systems that participate in intelligence itself.

This changes civilization fundamentally.

AI does not only affect tools.

It affects:

- truth,
- trust,
- employment,
- governance,

- identity,
- and human agency.

The scale of influence is planetary.

Why Existing Governance Is Insufficient

Current governance systems were designed for earlier eras.

They regulate:

- industries,
- markets,
- nations,
- technologies with limited reach.

Artificial Intelligence transcends these boundaries.

Today:

- algorithms influence global elections,
- digital platforms shape cultural narratives,
- automation transforms labor worldwide,
- AI decisions affect billions simultaneously.

Without shared guiding principles, intelligence evolves faster than civilization's ability to manage it.

The Risks of an Unchartered AI Civilization

Without a civilizational charter, AI development risks fragmentation.

1. Innovation Fragmentation
Organizations innovate independently without shared ethical alignment.

Result:

- inconsistent standards,
- unsafe deployments,
- technological inequality.

2. Governance Conflict
Nations regulate AI differently.

Result:

- regulatory competition,
- ethical inconsistency,
- geopolitical tension.

3. Declining Societal Trust
Citizens struggle to understand or trust intelligent systems influencing daily life.

Result:

- social polarization,
- resistance to innovation,
- institutional instability.

4. Unbounded Technological Power
Technological capability expands without moral direction.

Result:

- intelligence without accountability,
- efficiency without empathy,
- progress without sustainability.

The greatest risk of the AI era is not technological failure.

It is ethical absence.

Why Humanity Needs a New Charter

Artificial Intelligence requires a new form of agreement one extending beyond nations and organizations.

A civilization charter must answer fundamental questions:

- What responsibilities accompany intelligent power?
- How should AI serve humanity?
- How do we protect future generations?
- How do we sustain planetary balance while advancing innovation?

The AI era demands a charter not merely for governance but for civilization itself.

The ZHA Civilization Charter

The ZHA Civilization Charter emerges as a response to this need.

It establishes a common foundation for the Intelligent Age by aligning humanity around:

- Sustainable AI with Human-Centric Purpose,
- planetary stewardship,
- responsible innovation,
- shared prosperity,
- long-term civilization wellbeing.

Rather than restricting innovation, the Charter provides direction.

It transforms AI development from isolated technological activity into conscious civilization advancement.

Charter as Collective Responsibility

The ZHA Charter does not belong to a single nation, institution, or organization.

It represents a shared human commitment.

It invites participation from:

- students,
- professionals,
- industries,
- governments,
- researchers,
- communities worldwide.

Civilization becomes a cooperative project.

From Technological Power to Civilizational Wisdom

Humanity has learned to manage:

- land through governance,
- economies through institutions,
- societies through law.

Now humanity must learn to guide intelligence itself.

A charter marks this transition.

It signals that humanity has matured enough to recognize that intelligence must serve civilization not overwhelm it.

The ZHA Insight

Every civilization defines itself through its agreements.

The Industrial Age agreed to productivity.
The Digital Age agreed to connectivity.

The Intelligent Age must agree to responsibility.

The ZHA Civilization Charter establishes humanity's shared commitment:

to ensure intelligence advances with sustainability, ethics, and human purpose.

16.2 The Need for a Civilization Constitution in the AI Era

From National Governance to Governance of Intelligence

Throughout history, constitutions have been humanity's most powerful instruments for stability.

A constitution transforms ideals into operating systems for society.

It defines:

- authority and responsibility,
- rights and protections,
- decision-making boundaries,
- long-term societal direction.

Traditional constitutions govern nations.

They organize how people live together within geographical boundaries.

But the emergence of Artificial Intelligence introduces a new reality:

- intelligence itself now operates beyond borders.

Humanity has entered an era where governance must expand from nations to civilization.

The Limits of Nation-Based Constitutions

National constitutions were designed for a world defined by territory.

They regulate:

- governments,
- citizens,
- economies,
- legal systems.

However, Artificial Intelligence does not recognize geography.

AI systems:

- operate simultaneously across continents,
- learn continuously from global data,
- influence decisions instantly,
- shape behavior across cultures and societies.

An algorithm deployed in one country can affect billions worldwide within hours.

This creates a governance gap.

Technological power has become global.

Governance remains national.

Artificial Intelligence - A Borderless Force

AI differs fundamentally from earlier technologies.

It is not merely infrastructure.

It is decision infrastructure.

AI systems now influence:

- financial markets,
- medical decisions,
- employment opportunities,
- public information ecosystems,
- democratic processes,
- environmental management.

Unlike previous technologies, AI actively participates in intelligence formation.

Human civilization is no longer the sole decision-maker.

For the first time, humanity must govern systems that learn and evolve autonomously.

The Emerging Governance Crisis
Without a civilizational constitutional framework, AI development risks fragmentation.

Fragmented Ethics
Different regions define acceptable AI behavior differently.

Result:

- inconsistent human protections,
- ethical inequality across societies.

Accelerated Economic Impact

AI-driven automation reshapes global labor markets faster than policy adaptation.

Result:

- economic instability,
- social displacement,
- widening inequality.

Influence Without Accountability

Algorithms shape public opinion and decision-making without shared oversight.

Result:

- declining institutional trust,
- social polarization.

Technological Power Without Stewardship

Innovation advances faster than collective wisdom.

Result:
intelligence grows without coordinated responsibility.

The AI era therefore presents humanity with a new challenge:

- Who governs intelligence when intelligence operates globally?

Why Humanity Needs a Civilizational Constitution

A civilizational constitution extends governance beyond political borders.

It establishes shared principles guiding humanity's interaction with intelligent systems.

Such a constitution does not control innovation.

It aligns innovation with civilization survival.

It provides answers to fundamental questions:

- How should AI serve humanity?
- What responsibilities accompany intelligent power?
- How do we protect future generations?
- How do we sustain planetary balance while advancing intelligence?

The constitution becomes humanity's agreement on how intelligence must evolve.

The ZHA Civilization Constitution

The ZHA Civilization Constitution emerges as a complementary governance framework for the Intelligent Age.

It does not replace national laws or sovereignty.

Instead, it creates alignment across them.

It functions as a civilizational layer of guidance.

Constitutional Role of ZHA

The ZHA Constitution provides:

Ethical Alignment

Establishes universal principles ensuring AI respects:

- human dignity,
- fairness,

- transparency,
- accountability.

Organizations design systems responsibly before deployment.

Sustainability Principles

Ensures AI development protects:

- environmental systems,
- social wellbeing,
- intergenerational equity.

Innovation becomes sustainable by design.

Collaborative Governance Standards

Encourages cooperation between:

- governments,
- industries,
- academia,
- civil society.

Governance evolves from isolated regulation to shared stewardship.

Long-Term Civilization Stewardship

Introduces future-oriented decision-making.

Leaders evaluate:

- societal impact decades ahead,
- planetary consequences,
- cultural stability.

Technology aligns with civilization continuity.

Constitution as Humanity's Operating System for Intelligence

If the Charter defines civilization values, the Constitution operationalizes them.

The ZHA Civilization Constitution becomes:

- a framework guiding AI transformation,
- a reference for responsible innovation,
- a foundation for global collaboration,
- a stabilizing force for intelligent civilization.

It acts as an operating system governing humanity's relationship with intelligence itself.

Practical Example - Global AI Deployment

Consider a powerful generative AI system released worldwide.

Without civilizational constitutional guidance:

- companies prioritize speed,
- ethical standards vary,
- societal risks escalate.

Under ZHA constitutional alignment:

- sustainability impact assessed,
- governance safeguards applied,
- global collaboration ensured,
- long-term societal outcomes evaluated.

Innovation proceeds responsibly.

Civilization remains stable.

From Technology Governance to Intelligence Stewardship

Humanity has historically governed:

- land through law,
- economies through institutions,
- societies through constitutions.

The AI era requires governance of intelligence.

This represents humanity's next evolutionary step.

Governance shifts from controlling systems to stewarding intelligence responsibly.

The ZHA Insight

Artificial Intelligence is humanity's most powerful creation.

But power without shared constitutional guidance creates instability.

The ZHA Civilization Constitution ensures that intelligence evolves within a framework of:

- ethics,
- sustainability,
- collaboration,
- human-centered purpose.

It marks humanity's transition from technological civilization toward conscious intelligent civilization.

16.3 From National Governance to Civilization Governance

The Evolution of Governance in the Age of Intelligent Systems

Governance has always evolved alongside human capability.

As civilization gained new powers, humanity developed new governance structures to manage those powers responsibly.

Agriculture required local governance.
Industrialization required national governance.
Digital globalization required international coordination.

Artificial Intelligence now introduces a transformation that extends beyond all previous stages.

AI reshapes intelligence itself.

This shift demands a new form of governance:

Civilization Governance.

Governance Evolves with Human Capability

Human history reveals a clear progression.

Each technological era expanded human influence, forcing governance to expand accordingly.

Governance expands when human power expands.

Artificial Intelligence represents humanity's greatest expansion of power.

Therefore governance must evolve once again.

Stage One - Industrial Era Governance

Nation-Based Governance

The Industrial Revolution transformed societies through machines, factories, and mass production.

Economic activity became centralized within nations.

Governance focused on:

- national laws,
- labor protection,
- economic regulation,
- infrastructure development.

Nation-states became the primary governance units.

This model succeeded because industrial systems were geographically bound.

Factories existed within borders.

Production occurred locally.

National governance was sufficient.

Example - Industrial Governance

Governments introduced:

- labor rights,
- safety regulations,
- public education systems,
- national economic planning.

Industrialization stabilized only after governance matured.

Technology alone did not create civilization stability.

Governance did.

Stage Two - Digital Era Governance

Global Economic Governance

The Digital Revolution dissolved geographical boundaries.

Information began flowing globally.

Markets interconnected.

Digital platforms influenced billions simultaneously.

Governance expanded beyond nations through:

- international trade agreements,
- cybersecurity cooperation,
- global financial regulations,
- multinational institutional frameworks.

Economic globalization required coordination between nations.

Example - Digital Governance

Global institutions emerged to manage:

- ross-border commerce,
- internet standards,
- financial stability,
- data protection agreements.

Governance began adapting to interconnected humanity.

However, governance still focused primarily on economics and information exchange.

Stage Three - The AI Era

Civilization Governance

Artificial Intelligence introduces a fundamentally different challenge.

AI does not merely connect humanity.

It participates in decision-making itself.

AI influences:

- knowledge creation,
- social behavior,
- economic opportunity,
- environmental management,
- governance processes.

Intelligence becomes distributed across machines and humans.

At this stage, governance must expand beyond economics or nations.

It must govern civilization outcomes.

What Is Civilization Governance?

Civilization governance is the coordination of intelligence for collective human wellbeing.

It ensures that AI development serves humanity as a whole rather than isolated interests.

Civilization governance asks new questions:

- How does intelligence affect humanity long-term?
- What responsibilities accompany algorithmic power?
- How do we protect future generations?
- How do we sustain planetary ecosystems while advancing innovation?

Governance shifts from control to stewardship.

Why National Governance Alone Is No Longer Enough

AI systems operate globally the moment they are deployed.

A single algorithm can:

- influence global elections,
- reshape financial markets,
- alter cultural narratives,
- affect environmental systems worldwide.

When intelligence operates at planetary scale, governance limited to national interests becomes insufficient.

Without civilization governance:

- technological standards fragment,
- ethical protections vary,
- societal trust declines,
- instability increases.

Humanity requires shared governance aligned with collective survival.

The ZHA Model of Civilization Governance
ZHA introduces a governance framework rooted in four foundational principles.

1. Responsibility

Intelligence creators accept accountability for societal impact.

Organizations evaluate:

- environmental consequences,
- workforce effects,
- ethical risks,
- before deploying AI systems.

Responsibility becomes proactive rather than reactive.

2. Sustainability
AI development aligns with long-term planetary wellbeing.

Innovation considers:

- energy consumption,
- environmental preservation,
- intergenerational equity.

Progress becomes sustainable across decades, not quarters.

3. Inclusivity

Civilization governance ensures participation from all stakeholders:

- students,
- professionals,
- industries,
- governments,
- communities.

AI transformation becomes shared human participation rather than elite control.

Inclusivity builds trust.

4. Foresight
Civilization governance anticipates future impact.

Leaders ask:

- What will this technology mean in 20 years?
- How will society adapt?
- What unintended consequences may arise?

Governance evolves from reaction to anticipation.

Practical Example - AI Healthcare Governance.

Without civilization governance:

- AI healthcare systems optimize profit,
- data ownership conflicts arise,
- unequal access increases.

Under ZHA Civilization Governance:

- ethical data standards shared globally,
- accessibility prioritized,
- long-term societal health evaluated.

Technology strengthens civilization rather than fragmenting it.

Governance as Stewardship of Intelligence

Civilization governance represents humanity's maturity.

Humanity moves from:

- governing territory,
- governing markets,
- governing information,
- to governing intelligence itself.

Governance becomes stewardship of humanity's most powerful creation.

The ZHA Insight

The future of civilization will not be determined solely by technological breakthroughs.

It will be determined by how wisely humanity governs intelligence.

National governance stabilized industrial society.

Global economic governance stabilized digital society.

Civilization governance will stabilize the AI age.

ZHA provides the framework enabling this transition.

16.4 The ZHA Civilization Charter - Foundational Principles

Defining the Ethical Foundation of a Sustainable AI Civilization

Every enduring civilization is guided by shared principles that define not only how societies function, but why they exist.

Laws regulate behavior.
Policies guide operations.
But charters define civilization identity.

The ZHA Civilization Charter establishes the philosophical and ethical commitments required for humanity to thrive in the age of Artificial Intelligence.

It answers a fundamental civilizational question:

- How should intelligence evolve so that humanity flourishes alongside it?

The Charter transforms Sustainable AI from a technological initiative into a civilization philosophy one capable of guiding humanity across generations.

The Purpose of the ZHA Civilization Charter

Artificial Intelligence represents humanity's most powerful creation.

Without guiding principles, intelligence risks amplifying:

- inequality,
- environmental degradation,
- social fragmentation,
- unchecked technological dominance.

The ZHA Civilization Charter ensures that AI development aligns with human values, planetary wellbeing, and long-term civilization stability.

It establishes commitments that transcend:

- nations,
- industries,
- institutions,
- generations.

These commitments form the ethical architecture of a Human-Centric Intelligent Civilization.

Core Civilizational Commitments

1. Human-Centric Intelligence

Principle

Artificial Intelligence exists to enhance human dignity, capability, and wellbeing.

Technology must never reduce humanity to data points or economic variables.

Instead, AI should:

- empower human creativity,
- support meaningful work,
- improve quality of life,
- strengthen social cohesion.

Human beings remain the purpose of intelligence not its byproduct.

Example - Human-Centered Healthcare AI

Traditional AI optimization might prioritize efficiency alone.

Human-centric AI evaluates:

- patient trust,
- emotional wellbeing,
- accessibility,
- fairness of treatment.

AI assists doctors rather than replacing human compassion.

Outcome:
Technology strengthens humanity rather than diminishing it.

Civilizational Impact
Human identity remains central even in an intelligent technological world.

AI becomes an extension of human flourishing.

2. Planetary Stewardship

Principle

Technological progress must protect and regenerate the natural systems upon which civilization depends.

Human civilization exists within planetary limits.

AI must therefore become a tool of ecological balance.

Example - Sustainable AI Infrastructure

Instead of unlimited computational expansion:

- energy-efficient models are prioritized,
- renewable energy powers data centers,
- environmental impact is measured before deployment.

Innovation aligns with ecological responsibility.

Civilizational Impact

Humanity transitions from exploiting nature to collaborating with planetary systems.

Technology becomes a guardian of Earth rather than a threat to it.

3. Shared Prosperity

Principle

The benefits of intelligent technology must be distributed equitably across society.

AI must not concentrate power or opportunity within limited groups.

Innovation succeeds only when its benefits are shared.

Example - Inclusive AI Economy

Responsible AI deployment includes:

- workforce reskilling programs,
- accessible digital education,
- support for emerging economies,
- equitable access to intelligent tools.

Automation increases opportunity instead of displacement.

Civilizational Impact

Economic growth becomes inclusive rather than divisive.

Prosperity strengthens social stability.

4. Responsible Innovation

Principle

Innovation must consider long-term societal consequences before implementation.

The question shifts from:

- "Can we build this?"

to

- "Should we build this responsibly?"

Responsible innovation anticipates:

- ethical risks,
- social impact,
- environmental consequences,
- future generational effects.

Example - Autonomous Systems Deployment

Before deploying autonomous transportation:

- safety evaluated across diverse communities,
- employment transition plans developed,
- environmental benefits assessed.

Innovation proceeds with foresight.

Civilizational Impact

Progress becomes sustainable rather than disruptive.

Innovation aligns with wisdom.

5. Collective Collaboration

Principle

Humanity advances through cooperation rather than domination.

The AI era reveals humanity's interdependence.

Global challenges:

- climate change,
- digital security,
- health crises,

- information stability
- cannot be solved competitively.

Example - Global AI Safety Collaboration

Organizations share:

- alignment research,
- sustainability standards,
- ethical development practices.

Innovation accelerates while risks decline.

Collaboration becomes civilization strategy.

Civilizational Impact

Humanity recognizes a shared technological destiny.

Competition evolves into collaborative advancement.

6. Continuous Learning Civilization

Principle

Civilization must evolve through reflection, adaptation, and learning.

AI introduces continuous change.

Static societies cannot remain stable.

A Sustainable AI civilization embraces learning as permanent cultural practice.

Example - Adaptive Governance

Governments use AI-supported feedback systems to:

- evaluate policy outcomes,
- correct bias,
- update governance approaches dynamically.

Society learns alongside technology.

Civilizational Impact

Civilization becomes resilient.

Humanity evolves consciously rather than reactively.

From Sustainability Policy to Civilizational Identity

Traditional sustainability exists as compliance:

- reports,
- audits,
- regulatory requirements.

The ZHA Civilization Charter transforms sustainability into identity.

Organizations and societies begin asking:

- Is this innovation aligned with human dignity?
- Does it protect the planet?
- Does it benefit future generations?

Sustainability becomes embedded behavior.

Ethics becomes culture.

Responsibility becomes instinct.

The ZHA Insight

Civilizations endure not because of technological strength, but because of moral clarity.

The ZHA Civilization Charter establishes the ethical compass required for humanity's intelligent future.

It ensures that:

- intelligence remains human-centered,
- innovation remains sustainable,
- progress remains inclusive,
- civilization remains conscious.

16.5 The ZHA Civilization Constitution - Operational Framework

From Civilizational Values to Living Governance Systems

Civilizations endure when ideals become systems.

The ZHA Civilization Charter defines humanity's shared values for the Intelligent Age.

The ZHA Civilization Constitution ensures those values become daily practice.

If the Charter answers why civilization must guide intelligence, the Constitution answers how civilization governs intelligence responsibly.

It operationalizes Sustainable AI with Human-Centric Purpose through institutional mechanisms capable of functioning across:

- education,
- industry,
- governance,

- research,
- global collaboration.

The Constitution transforms Sustainable AI from philosophy into actionable civilization governance.

The Role of a Civilization Constitution

Unlike national constitutions that regulate governments, the ZHA Civilization Constitution guides the interaction between:

- humans,
- institutions,
- intelligent systems,
- planetary ecosystems.

Its purpose is not control.

Its purpose is alignment.

It ensures that intelligence evolves responsibly while innovation continues to flourish.

The Constitution creates a stable operating environment for intelligent civilization.

Constitutional Pillars of Sustainable AI Civilization

1. Ethical Intelligence Governance

Purpose

To ensure Artificial Intelligence remains accountable, transparent, and aligned with human values.

As AI systems increasingly participate in decision-making, governance must shift from oversight after deployment to responsibility embedded within design.

Key Mechanisms

AI Transparency Requirements

Organizations must disclose:

- how AI systems make decisions,
- what data influences outcomes,
- limitations and risks of algorithms.

Transparency builds public trust.

Example:
A bank deploying AI loan approval systems publishes explainable decision models so customers understand outcomes.

Algorithm Accountability

Creators remain responsible for AI behavior even after deployment.

Accountability includes:

- auditability,
- traceability,
- human oversight mechanisms.

Example:
Healthcare AI recommendations always require human validation before medical decisions.

Bias Mitigation Systems

AI must actively prevent discrimination.

Organizations implement:

- fairness testing,
- inclusive dataset governance,
- continuous monitoring.

Example:
Hiring AI systems tested across gender, socioeconomic, and cultural datasets before release.

Ethical Review Councils

Independent multi-disciplinary councils evaluate AI deployments.

Participants include:

- technologists,
- ethicists,
- sustainability experts,
- community representatives.

Ethics becomes institutionalized rather than optional.

Civilizational Outcome

Intelligence becomes trustworthy.

Governance evolves from regulation → stewardship.

2. Sustainable Innovation Standards

Purpose

To align technological innovation with Environmental, Social, and Governance (ESG) sustainability principles.

Innovation must not sacrifice long-term civilization health for short-term advantage.

Key Mechanisms

ESG-Aligned Product Development

Every AI product undergoes sustainability evaluation:

- environmental footprint,
- workforce impact,
- governance implications.

Example:
An AI logistics platform optimized not only for speed but also carbon emission reduction.

Energy-Efficient AI Design

Organizations prioritize:

- optimized model architectures,
- low-energy computing,
- sustainable infrastructure.

Example:
Choosing efficient models over larger resource-intensive models when outcomes are equivalent.

Lifecycle Sustainability Evaluation

Technology assessed across full lifecycle:

- development,
- deployment,

- operation,
- retirement.

Innovation becomes responsible from creation to decommissioning.

Civilizational Outcome

Progress becomes regenerative rather than extractive.

Technology supports planetary continuity.

3. Education for Civilization Stewardship

Purpose

To prepare humanity to live responsibly within an intelligent civilization.

Technology alone cannot sustain civilization.

People must evolve alongside intelligence.

Key Mechanisms

Sustainability Mindset Curriculum

Education integrates:

- ethics,
- environmental awareness,
- social responsibility,
- human-centered innovation.

Programs like ZSPC develop Sustainability Mindset Practitioners.

Responsible Innovation Training

Professionals learn:

- ethical product development,
- AI ESG integration,
- long-term decision-making.

Innovation becomes conscious practice.

Lifelong Learning Ecosystems

Learning continues throughout life through AI-assisted education systems.

Citizens adapt continuously to technological change.

Civilizational Outcome

Humanity becomes capable of guiding intelligence wisely.

Education evolves from knowledge transfer → civilization preparation.

4. Global Collaboration Structures

Purpose

To replace fragmented technological progress with coordinated global advancement.

AI challenges are planetary.

Solutions must be collaborative.

Key Mechanisms

Cross-Country Research Alliances

Universities and research centers collaborate on:

- sustainable AI architectures,
- climate intelligence,
- ethical algorithm development.

Knowledge becomes shared global infrastructure.

Shared Governance Standards

Common principles guide AI deployment worldwide.

Organizations align around:

- ethical norms,
- sustainability benchmarks,
- accountability frameworks.

Cooperative Technological Development

Industries collaborate on solving global problems instead of competing destructively.

Example:
Companies jointly develop climate prediction AI rather than proprietary isolated solutions.

Civilizational Outcome

Innovation accelerates while risks decrease.

Humanity acts as one intelligent ecosystem.

5. Civilization Impact Measurement

Purpose

To redefine success beyond economic output toward civilization wellbeing.

What civilization measures determines what civilization values.

Key Mechanisms

Wellbeing Indicators

Measurement expands to include:

- mental health,
- social cohesion,
- education accessibility,
- community resilience.

Sustainability Metrics

Organizations track:

- carbon footprint,
- resource efficiency,
- ecological regeneration.

Ethical Innovation Indices

Innovation evaluated based on:

- fairness,
- transparency,
- societal benefit,
- long-term impact.

Civilizational Outcome

Success shifts from GDP growth to Civilization Health.

Progress becomes meaningful.

Constitution as a Living Governance System

The ZHA Civilization Constitution is not a static legal document.

It is a living operational framework evolving alongside intelligence itself.

It allows:

- continuous learning,
- adaptive governance,
- participatory evolution.

Civilization becomes dynamic yet stable.

From Philosophy to Actionable Civilization Governance

Through its five pillars, the Constitution transforms Sustainable AI into everyday practice:

The Constitution makes Sustainable AI implementable across the world.

The ZHA Insight

Humanity has mastered innovation.

The next challenge is operationalizing wisdom.

The ZHA Civilization Constitution ensures that intelligence advances within structures capable of protecting:

- humanity,
- society,
- ecosystems,
- future generations.

It marks civilization's transition from inventing intelligence to governing intelligence responsibly.

16.6 Institutional Structures Supporting the Charter and Constitution

Operationalizing Sustainable AI Civilization Through Institutional Intelligence

A charter defines purpose.
A constitution defines governance.

But civilization becomes sustainable only when institutions embody those principles in action.

The ZHA Civilization Charter and Constitution are not theoretical declarations.

They operate through an interconnected ecosystem of institutions designed to guide humanity's relationship with Artificial Intelligence responsibly.

These institutions collectively form the execution architecture of Sustainable AI with Human-Centric Purpose.

Each institution fulfills a distinct role, yet all operate as parts of a unified civilizational system.

Together they transform Sustainable AI from ideology into global practice.

The Need for Institutional Intelligence

Throughout history, enduring civilizations depended on institutions capable of sustaining progress across generations.

Institutions:

- preserve knowledge,
- coordinate collective action,

- maintain ethical continuity,
- stabilize societal transformation.

Artificial Intelligence introduces challenges that no single organization or government can manage alone.

The ZHA institutional framework therefore acts as distributed civilization intelligence — coordinating education, governance, research, industry, and society.

Core Institutional Pillars of the ZHA Civilization Framework

1. ZHA Global Institute

The Intellectual and Governance Nucleus

The ZHA Global Institute serves as the central knowledge and coordination hub for Sustainable AI civilization.

It functions as the brain of the ZHA ecosystem, translating charter principles into practical global initiatives.

Primary Functions
Policy Development

The Institute develops frameworks guiding:

- Sustainable AI governance,
- ESG-aligned innovation standards,
- ethical technology deployment,
- global sustainability coordination.

Policies focus on foresight rather than reaction.

Example:
Developing Sustainable AI SDLC frameworks adopted by educational

institutions and industries worldwide.

Leadership Education

The Institute prepares leaders capable of navigating intelligent civilization challenges.

Programs cultivate:

- systems thinking,
- ethical decision-making,
- sustainability stewardship,
- human-centered innovation leadership.

Leaders evolve from managers into civilization stewards.

Global Coordination

The Institute connects governments, academia, industries, and communities into a shared ecosystem.

It facilitates:

- international collaboration,
- knowledge exchange,
- shared governance alignment.

The Institute becomes a coordinating intelligence layer for humanity's AI transformation.

Civilizational Role

The ZHA Global Institute ensures Sustainable AI remains guided by knowledge, ethics, and foresight.

2. ZHA World Council

Civilization-Scale Leadership and Foresight Body

The ZHA World Council represents humanity's collective leadership mechanism for the Intelligent Age.

Where the Institute develops knowledge, the Council guides direction.

Responsibilities
Ethical AI Direction

The Council evaluates emerging technologies against civilizational impact criteria:

- societal stability,
- ethical implications,
- environmental consequences.

It provides guidance before risks escalate.

Global Collaboration Facilitation

The Council brings together:

- policymakers,
- industry leaders,
- educators,
- sustainability practitioners.

It encourages cooperation across nations rather than technological rivalry.

Long-Term Civilization Planning

Unlike traditional leadership focused on short-term outcomes, the Council operates with multi
generational foresight.

Planning horizons include:

- 20-year technological trajectories,
- future workforce evolution,
- planetary sustainability challenges.

Example
Before widespread deployment of autonomous governance AI, the Council conducts global consultation assessing social implications.

Governance becomes anticipatory rather than reactive.

Civilizational Role

The ZHA World Council functions as humanity's ethical compass for intelligent civilization.

3. ZHA Sustainable AI Research Network

Global Collaboration for Civilization Intelligence

Research becomes civilization infrastructure under ZHA.

The Sustainable AI Research Network connects:

- universities,
- research laboratories,
- innovation centers,
- organizations
- industry technology teams.

Knowledge ceases to exist in isolation.

Core Focus Areas

Sustainable AI Architectures

Designing intelligent systems optimized for:

- energy efficiency,
- environmental sustainability,
- long-term operational stability.

Climate Intelligence

AI models support:

- climate prediction,
- biodiversity protection,
- resource optimization,
- disaster prevention.

AI becomes planetary guardian.

Societal Impact Modeling

Researchers study:

- workforce transformation,
- social equity outcomes,
- governance implications,
- human wellbeing metrics.

Technology development integrates social science insight.

Example
Researchers across continents jointly develop low-energy AI models reducing global data-center emissions.

Innovation becomes collective responsibility.

Civilizational Role

The Research Network transforms knowledge into civilization intelligence guiding humanity's future.

4. ZHA Education & Certification Ecosystem (ZSPC)

Developing Sustainability Practitioners Across Generations

Civilization transformation begins with people.

The ZHA Education & Certification Ecosystem builds a global population capable of guiding intelligent systems responsibly.

Core Functions

Sustainability Mindset Development

Students learn:

- ethical innovation,
- environmental stewardship,
- collaborative leadership,
- responsible technology use.

Education shifts from skill acquisition to mindset transformation.

Certification Pathways

Participants progress through structured sustainability certifications integrating:

- real-world projects,
- community engagement,
- AI awareness,
- governance literacy.

Multi-Generational Learning

Programs serve:

- school students,
- university learners,
- professionals,
- institutional leaders.

Learning becomes lifelong.

Outcome

A globally aligned workforce emerges - not merely AI users but Sustainability Mindset Practitioners.

Civilizational Role

ZSPC ensures humanity evolves alongside intelligence.

Education becomes civilization preparation.

5. ZHA Industry & Innovation Consortium

Industry as Steward of Civilization Progress

Industry holds enormous influence over technological direction.

The Consortium aligns corporations toward Sustainable AI transformation.

Industry moves from market dominance to civilization stewardship.

Key Activities

Responsible Innovation Alignment

Organizations integrate:

- ESG principles,

- ethical SDLC methodologies,
- human-centered design practices.

Collaborative Innovation Platforms

Companies share:

- sustainability standards,
- ethical frameworks,
- best practices.

Competition evolves into cooperative advancement.

Sustainable AI Deployment

Corporations deploy AI systems prioritizing:

- social wellbeing,
- environmental sustainability,
- long-term trust.

Example
Technology firms collaborate on climate intelligence AI instead of competing through proprietary systems.

Economic success aligns with planetary protection.

Civilizational Role

Industry becomes a partner in civilization continuity rather than a driver of disruption.

Institutional Pillars Integration - The ZHA Civilization Operating Model

These institutions function as an integrated ecosystem:

Together they operationalize the Charter and Constitution.

From Institution Pillars to Civilization Infrastructure

The ZHA institutional framework represents a new category of organization:

- not national,
- not corporate,
- not academic alone.

It is civilizational infrastructure.

It enables humanity to:

- coordinate intelligence,
- sustain innovation,
- protect planetary systems,
- evolve consciously.

The ZHA Insight

Technological revolutions succeed when supported by institutions capable of guiding them wisely.

Artificial Intelligence demands institutions designed specifically for responsible intelligence governance.

The ZHA institutional ecosystem ensures that Sustainable AI with Human-Centric Purpose becomes:

- teachable,
- governable,
- scalable,
- sustainable.

16.7 Collaborative Leadership as Constitutional Practice

Leadership under the ZHA Constitution shifts from authority to stewardship.

Leaders become:

- guardians of future generations,
- protectors of societal stability,
- architects of sustainable progress.

Decision-making expands beyond organizational benefit toward civilization impact.

16.8 Governance Through Participation

From Centralized Control to Collaborative Intelligence

For most of human history, governance has been hierarchical.

Authority flowed from:

- rulers,
- governments,
- institutions,
- centralized decision-making bodies.

Citizens participated indirectly through representation.

This model worked when societies changed slowly and decision cycles extended across decades.

The Artificial Intelligence era fundamentally changes this reality. AI accelerates change faster than centralized governance structures can respond.

Policies written today may become obsolete tomorrow.
Technologies evolve continuously.
Societal impact emerges in real time.

Under these conditions, governance must evolve. The ZHA Civilization Constitution introduces a new paradigm:

- Governance Through Participation.
- Governance becomes a living system shaped by collective human intelligence.

Why Traditional Governance Models Are Insufficient

Centralized governance assumes:

- information flows upward slowly,
- experts hold most knowledge,
- decisions can be made periodically.

Artificial Intelligence reverses these assumptions.

Today:

- knowledge is distributed,
- innovation happens everywhere,
- societal feedback occurs instantly,
- technological impact emerges continuously.

No single institution can fully understand the consequences of intelligent systems.

Governance must therefore become distributed.

Distributed Governance in the ZHA Model

The ZHA Constitution promotes governance that is shared across society.

Instead of concentrating authority, it coordinates participation across multiple actors.

Governance evolves from command-and-control toward collaborative intelligence.

Participation becomes a responsibility of civilization itself.

Participants in ZHA Distributed Governance

1. Citizens - Contributors of Collective Insight
Citizens are no longer passive recipients of policy.

They become active participants shaping intelligent civilization.

Role of Citizens:

- provide societal feedback,
- identify ethical concerns,
- contribute local knowledge,
- participate in sustainability initiatives.

AI-enabled platforms allow citizens to express informed perspectives on technological impact.

Example - Participatory Urban Governance

A smart city uses AI systems to gather citizen input on:

- transportation planning,
- environmental monitoring,
- public safety policies.

Community insights guide decision-making.

Governance reflects lived experience rather than abstract planning.

Civilizational Outcome

Democracy evolves from periodic voting to continuous participation.

Citizens become co-architects of civilization.

2. Educators - Shapers of Civilization Values

Education determines how future generations think about technology.

Educators play a governance role by shaping cultural foundations.

Role of Educators:

- teach sustainability mindset,
- cultivate ethical reasoning,
- develop responsible innovators,
- embed long-term thinking into learning systems.

Governance begins in classrooms.

Example - Sustainability Mindset Education (ZSPC)

Students engage in:

- environmental projects,
- community collaboration,
- ethical technology discussions.

Graduates enter society already aligned with responsible intelligence practices.

Governance becomes cultural rather than regulatory.

Civilizational Outcome

Values guide behavior even without supervision.

Ethical governance becomes instinctive.

3. Industries - Implementers of Responsible Standards

Industry builds the technologies shaping civilization.

Therefore, corporations become governance actors.

Role of Industry:

- implement Sustainable AI standards,
- integrate ESG into product development,
- maintain transparency and accountability,
- collaborate across sectors.

Industry moves beyond profit optimization toward civilization stewardship.

Example - Responsible AI Product Development

A technology company adopting ZHA principles:

- evaluates societal impact before release,
- monitors environmental footprint,
- maintains ethical audit trails.

Corporate decisions support civilization stability.

Civilizational Outcome

Innovation aligns with societal wellbeing.

Industry becomes partner in governance rather than subject of regulation alone.

4. Governments - Aligners of Policy Frameworks

Governments remain essential.

However, their role evolves.

Instead of controlling innovation after harm occurs, governments coordinate and align collective participation.

Role of Governments:

- establish enabling policies,
- support collaboration platforms,
- harmonize international standards,
- protect public interest.

Governments transition from regulators to stewards of intelligent society.

Example - AI Governance Alignment

Governments collaborate with:

- universities,
- industry consortia,
- citizen councils,
- global institutions.

Policies reflect collective intelligence rather than isolated authority.

Civilizational Outcome

Governance becomes adaptive, responsive, and trusted.

Governance as Collaborative Intelligence

When citizens, educators, industry, and governments participate together, governance evolves into a new form:

Collaborative Intelligence Governance.

Decision-making integrates:

- societal wisdom,
- scientific knowledge,
- technological expertise,
- ethical reflection.

Governance becomes dynamic rather than static.

AI as an Enabler of Participatory Governance

Artificial Intelligence itself enables this transformation.

AI systems support:

- real-time societal feedback analysis,
- transparent policy simulations,
- inclusive decision platforms,
- impact forecasting.

Technology strengthens democracy rather than replacing it.

AI becomes governance infrastructure.

The Cultural Shift Behind Participatory Governance

Governance through participation requires cultural evolution.

Humanity shifts from:

- dependency → responsibility,
- authority → collaboration,
- compliance → contribution.

Citizens begin seeing governance as shared stewardship of civilization.

Real-World Scenario - Distributed Sustainable AI Governance

Under ZHA alignment:

Students learn sustainability mindset.

Professionals implement responsible innovation.

Industries adopt ESG-driven SDLC practices.

Citizens contribute insight.

Governments coordinate global standards.

Outcome:
Civilization governs intelligence collectively.

Stability emerges naturally.

The ZHA Insight

The greatest governance challenge of the AI era is not lack of intelligence.

It is lack of participation. Civilization becomes stable when governance reflects collective wisdom rather than centralized power.

The ZHA Constitution transforms governance into a participatory ecosystem where humanity guides intelligence together.

16.9 Research as Constitutional Intelligence

From Academic Knowledge to Civilization Guidance

Throughout history, research has been humanity's primary mechanism for discovery.

Universities produced knowledge.
Scientists advanced understanding.
Innovation emerged from experimentation and inquiry.

Yet in most modern systems, research has largely remained confined to:

- academic publications,
- specialized disciplines,
- institutional recognition,
- technological advancement without societal integration.

The Artificial Intelligence era fundamentally changes the purpose of research.

When intelligence itself becomes programmable, research is no longer merely intellectual exploration.

It becomes a civilizational responsibility.

Under the ZHA Civilization Constitution, research evolves into Constitutional Intelligence, the thinking system that continuously guides humanity's relationship with technology, society, and the planet.

Why Research Must Evolve in the AI Era

Artificial Intelligence accelerates change faster than any previous technological revolution.

AI systems:

- learn continuously,
- reshape economies,

- influence culture,
- alter governance structures,
- impact environmental systems.

Civilization cannot rely solely on reactive problem-solving.

It must anticipate consequences before disruption occurs.

Research therefore becomes essential for:

- foresight,
- ethical alignment,
- sustainability planning,
- civilization stability.

Research moves from curiosity-driven inquiry toward responsibility-driven intelligence.

Research as a Constitutional Function

In the ZHA framework, research serves the same purpose for civilization that a constitution serves for governance.

A constitution defines principles.

Research continuously interprets and applies those principles to emerging realities.

Constitutional intelligence ensures that civilization adapts wisely as technology evolves.

Research becomes an ongoing dialogue between:

- innovation,
- ethics,
- society,

- and sustainability.

Core Domains of ZHA Constitutional Research

1. Sustainable AI Design

Purpose

To ensure intelligent systems evolve within planetary and societal limits.

Traditional AI research prioritized performance:

- larger models,
- faster computation,
- increased automation.

ZHA-guided research introduces sustainability as a primary design objective.

Research Focus:

- energy-efficient machine learning,
- low-resource AI architectures,
- carbon-aware computing,
- circular lifecycle technology design.

Example - Energy-Efficient Intelligence

Researchers redesign AI models to achieve equivalent performance using significantly reduced computational energy.

Impact:

- reduced environmental footprint,
- scalable global deployment,

- sustainable technological growth.

Civilizational Outcome

Innovation no longer conflicts with environmental survival.

Technology becomes ecologically aligned.

2. Human-AI Collaboration Models

Purpose

To redefine the relationship between humans and intelligent systems.

The risk of AI adoption lies not in automation itself but in poorly designed interaction between humans and machines.

ZHA research explores how intelligence can augment humanity rather than replace it.

Research Focus:

- human-in-the-loop systems,
- augmentation-based workflows,
- cognitive collaboration design,
- ethical automation models.

Example - Augmented Healthcare Decision Systems

AI analyzes medical data while physicians maintain final judgment.

Human empathy combines with machine precision.

Outcome:
Better decisions without loss of human agency.

Civilizational Outcome

AI strengthens human capability and purpose.

Human dignity remains central to technological progress.

3. Ethical Intelligence Systems

Purpose

To embed ethics directly into intelligent systems rather than applying ethical correction after deployment.

Ethics becomes design architecture.

Research Focus:

- bias detection methodologies,
- explainable AI frameworks,
- value-aligned machine learning,
- accountability modeling.

Example - Ethical Recommendation Algorithms

Research teams develop AI systems capable of identifying harmful misinformation patterns while preserving freedom of expression.

Technology balances openness with responsibility.

Civilizational Outcome

Trust becomes foundational to intelligent civilization.

Ethics transitions from regulation to system behavior.

4. Environmental Resilience Technologies

Purpose

To use intelligence for planetary protection and long-term ecological stability.

Climate change and environmental stress represent civilization-scale challenges requiring intelligent coordination.

Research Focus:

- climate prediction systems,
- biodiversity monitoring AI,
- sustainable agriculture intelligence,
- disaster resilience modeling.

Example - Climate Intelligence Platforms

Global AI systems analyze environmental data to forecast droughts, floods, and ecosystem changes.

Governments and communities prepare proactively.

Impact:
Technology prevents crises rather than reacting to them.

Civilizational Outcome

Human civilization becomes adaptive and resilient within planetary boundaries.

Knowledge as Civilization Guidance

Under ZHA alignment, knowledge ceases to be isolated academic output.

Research outcomes must guide:

- policy formation,

- industrial innovation,
- education systems,
- governance decision-making.

Research becomes actionable intelligence.

Academia, industry, and governance integrate into a shared learning ecosystem.

From Research Institutions to Civilization Intelligence Networks

The ZHA Sustainable AI Research Network transforms research into a collaborative global system.

Instead of isolated discovery:

- universities collaborate internationally,
- industries share sustainability insights,
- policymakers engage researchers continuously.

Knowledge flows into civilization governance in real time.

Example - Constitutional Intelligence in Practice

A new AI technology emerges with potential societal risk.

Under traditional systems:
Research follows deployment.

Under ZHA Constitutional Intelligence:

- research anticipates impact,
- ethical evaluation precedes release,
- sustainability assessment guides design,
- governance adapts proactively.

Research becomes civilization foresight.

The Cultural Transformation of Research

Researchers evolve from:
knowledge specialists
to
civilization stewards.

Scientific inquiry gains ethical purpose.

Innovation aligns with humanity's long-term survival.

Research culture shifts toward:

- interdisciplinarity,
- collaboration,
- societal responsibility,
- planetary awareness.

The ZHA Insight

Civilizations survive not because they innovate faster, but because they learn faster.

Research becomes humanity's mechanism for continuous self-correction.

The ZHA Civilization Constitution elevates research into Constitutional Intelligence — the guiding mind of Sustainable AI civilization.

16.10 Scaling Sustainable AI Globally

From Enforcement to Alignment: The Natural Expansion of Intelligent Civilization

Historically, large-scale societal change has often relied on enforcement mechanisms:

- laws imposed from authority,
- regulations mandating compliance,
- institutional directives controlling behavior.

While enforcement creates order, it rarely creates lasting transformation.

Civilizations endure not because people are compelled to change, but because they choose to align with shared values.

The ZHA Civilization Charter and Constitution introduce a fundamentally different approach to global transformation:

Sustainable AI scales through alignment rather than enforcement.

Instead of imposing uniform control, the ZHA model enables organic adoption rooted in shared purpose, collective responsibility, and mutual benefit.

Civilization evolves naturally when humanity recognizes a common intelligent future.

Why Enforcement Alone Cannot Scale Sustainable AI

Artificial Intelligence operates within decentralized systems:

- global technology networks,
- diverse cultures,
- independent institutions,
- sovereign nations.

No single authority can govern global intelligence entirely.

Attempts at rigid control often produce:

- innovation resistance,
- regulatory fragmentation,
- technological inequality,
- reduced global trust.

The AI era therefore requires a governance model capable of spreading through cooperation rather than coercion.

ZHA introduces alignment-based scaling.

The ZHA Model of Organic Civilization Expansion

Under the ZHA framework, Sustainable AI spreads through interconnected adoption pathways.

Each societal domain becomes both participant and catalyst for transformation.

Civilization evolves as values propagate across systems.

Pathways for Global Scaling

1. Educational Institutions — Seeding Civilization Mindset
Education represents the starting point of long-term civilization transformation.

When schools and universities integrate sustainability mindset education, future generations naturally adopt responsible innovation practices.

Transformation in Education

Institutions embed:

- sustainability thinking,
- ethical AI awareness,
- collaborative problem-solving,
- long-term societal responsibility.

Programs such as ZSPC cultivate students as Sustainability Mindset Practitioners.

Example
Students trained in sustainability mindset enter industries already aligned with responsible AI principles.

Organizations do not need to enforce ethics - employees carry them inherently.

Scaling Effect

Education creates generational continuity.

Transformation becomes self-sustaining.

2. Organizations - Implementing Responsible SDLC Practices

Organizations translate principles into operational reality.

Under ZHA alignment, enterprises adopt Sustainable AI Software Development Life Cycles (SDLC).

Responsible SDLC integrates:

- ESG impact evaluation,
- ethical risk assessment,
- human-centered design,
- sustainability metrics.

Example - Sustainable Product Development

A technology company develops an AI platform considering:

- energy consumption,
- workforce implications,

- long-term societal outcomes.

Responsible innovation becomes competitive advantage.

Scaling Effect

Industry adoption spreads through demonstrated success rather than regulation.

Responsible innovation becomes the global norm.

3. Governments - Aligning ESG Governance Frameworks

Governments play a coordinating role rather than an enforcement-only role.

Under ZHA principles, policymakers align national frameworks with Sustainable AI values.

Governments:

- encourage ESG-integrated innovation,
- support sustainability education,
- enable collaborative research,
- harmonize international AI standards.

Example
Multiple countries adopt shared ethical AI principles inspired by ZHA governance models.

Policy alignment reduces global fragmentation.

Scaling Effect

Governance evolves collaboratively across borders.

Civilization stability increases.

4. Communities - Participatory Technological Governance

Civilization transformation succeeds only when communities participate.

Citizens engage actively in shaping intelligent systems affecting their lives.

Participation includes:

- community sustainability initiatives,
- local innovation ecosystems,
- citizen feedback platforms,
- participatory digital governance.

Example - Community AI Governance

Local communities use AI platforms to monitor environmental quality and influence municipal policy decisions.

Technology strengthens democratic participation.

Scaling Effect

Governance becomes culturally embedded.

Society transitions from passive technology consumption to active stewardship.

Alignment as the Engine of Civilization Evolution

The ZHA Charter and Constitution function like a shared cultural compass.

Organizations and societies adopt principles because alignment produces measurable benefits:

- trust increases,
- innovation stabilizes,

- sustainability improves,
- collaboration accelerates.

Transformation spreads through inspiration and demonstrated success.

Not enforcement.

Network Effects of Sustainable AI Adoption

As adoption grows, a global network effect emerges:

- Educated citizens demand responsible technology.
- Organizations respond with Sustainable AI practices.
- Governments align governance frameworks.
- Communities reinforce cultural adoption.
- Each participant strengthens the system.
- Civilization evolves through interconnected learning.

Organic Civilization Evolution

Unlike industrial revolutions driven by centralized authority, Sustainable AI civilization evolves organically.

Shared values act as invisible infrastructure connecting humanity.

Civilization transformation becomes:

- adaptive,
- inclusive,
- resilient,
- self-reinforcing.

Humanity moves from fragmented progress toward coordinated evolution.

Real-World Scenario - Global Sustainable AI Scaling

Imagine global adoption under ZHA alignment:

- universities teach sustainability mindset worldwide,
- corporations deploy ethical AI as industry standard,
- governments cooperate on ESG governance,
- research institutions collaborate internationally,
- citizens participate in technological decision-making.

Result:
Sustainable AI becomes civilization default behavior.

No single entity enforces it.

Humanity collectively sustains it.

The ZHA Insight

Civilizations cannot be engineered through control.

They grow through shared belief and aligned purpose.

The ZHA Civilization Charter and Constitution succeed because they offer humanity a vision worth adopting:

Sustainable AI with Human-Centric Purpose.

When people align with purpose, transformation becomes inevitable.

16.11 The Civilization Constitution in Practice - A Living Governance Scenario

From Principle to Practice

A constitution becomes meaningful only when it moves beyond written ideals and begins shaping everyday decisions.

The ZHA Civilization Constitution is not intended to remain a philosophical declaration.
It is designed to function as living governance - a dynamic system guiding how nations educate citizens, develop technology, govern institutions, and cooperate globally.

This section presents a practical scenario:

How a nation transforms when it adopts the ZHA Civilization Charter and Sustainable AI principles.

The Starting Point - A Nation at Technological Crossroads

Consider a nation entering rapid AI transformation.

The country faces familiar global challenges:

- accelerating automation,
- economic uncertainty,
- environmental stress,
- declining public trust in institutions,
- fragmented technological adoption.

AI adoption begins quickly across sectors, but leaders recognize a fundamental risk:

Technology without civilizational alignment may create instability rather than prosperity.

The nation therefore adopts the ZHA Civilization Charter as its guiding framework.

Transformation begins across four interconnected domains.

1. Sustainability Mindset Education in Schools
Education as the Constitutional Foundation

The first implementation step recognizes a central ZHA insight:

Civilization governance begins with human mindset, not policy enforcement.

Education systems integrate sustainability mindset development from early schooling through higher education.

Educational Transformation Includes:

- sustainability ethics integrated into curricula,
- AI literacy combined with social responsibility,
- project-based community sustainability programs,
- interdisciplinary learning linking technology, environment, and society.

Students learn not only how to use technology but how to guide it responsibly.

Example - Classroom Transformation

Students participate in national sustainability certification programs:

- designing local climate solutions,
- analyzing AI bias scenarios,
- developing community digital governance projects,
- measuring environmental impact through data analytics.

Education produces responsible innovators, not passive technology consumers.

Constitutional Outcome

Future generations naturally uphold the Constitution's values.

Governance becomes cultural rather than regulatory.

2. ESG-Aligned AI Industry Policies

Industry as Civilization Builder

Under the ZHA Constitution, economic growth and sustainability are no longer competing priorities.

National AI policy mandates alignment with ESG principles.

Organizations adopting AI must demonstrate:

- environmental efficiency,
- social responsibility,
- transparent governance.

Policy Implementation

Government introduces:

- Sustainable AI certification standards,
- energy-efficient computing incentives,
- workforce transition programs,
- ethical AI review boards.

Companies innovate freely but within a shared sustainability framework.

Example - Manufacturing Sector

An AI-enabled manufacturing ecosystem adopts:

- predictive maintenance reducing waste,
- energy-optimized production systems,
- worker reskilling instead of layoffs,
- circular supply chains minimizing environmental impact.

Profitability increases while environmental footprint decreases.

Constitutional Outcome
Economic resilience strengthens because innovation becomes sustainable by design.

3. Transparent Governance Supported by AI

Governance Evolves into Intelligent Stewardship

The ZHA Constitution transforms governance from bureaucratic administration into transparent, data-informed stewardship.

AI supports governance but does not replace human accountability.

Governance Transformation Includes:

- AI-assisted public policy simulations,
- real-time environmental monitoring,
- transparent public expenditure tracking,
- participatory citizen feedback platforms.

Government decisions become explainable and evidence-based.

Example - Public Policy Decision System

Before approving a large infrastructure project:

AI models simulate:

- environmental impact,
- employment effects,
- urban sustainability outcomes,
- long-term social consequences.

Citizens access open dashboards showing decision rationale.

Constitutional Outcome
Trust between citizens and institutions increases dramatically.

Governance becomes collaborative rather than authoritative.

4. Collaborative International Research Partnerships

Civilization Governance Extends Beyond Borders

The ZHA Constitution recognizes that intelligent systems operate globally.

Therefore, national progress depends on international cooperation.

The nation participates in ZHA Global Institute networks connecting:

- universities,
- research labs,
- sustainability organizations,
- technology innovators.

Example - Global Climate Intelligence Alliance

Multiple countries share AI models predicting:

- climate risks,
- agricultural sustainability,
- disaster response strategies.

Knowledge becomes a shared global resource rather than proprietary competition.

Constitutional Outcome

Global collaboration replaces technological isolation.

Civilization resilience strengthens collectively.

Integrated Outcomes of Living Constitutional Governance

When the ZHA Civilization Constitution operates across education, industry, governance, and research, measurable societal outcomes emerge.

Economic Resilience

- sustainable innovation attracts long-term investment,
- workforce adapts continuously through reskilling,
- industries remain competitive without social disruption.

Economic growth becomes stable rather than volatile.

Environmental Protection

AI systems actively protect ecosystems through:

- energy optimization,
- climate intelligence,
- resource conservation.

Technology becomes a guardian of the planet.

Increased Social Trust
Transparent governance and ethical innovation restore public confidence.

Citizens feel:

- included,
- protected,
- empowered.

Trust becomes national capital.

Ethical Technological Growth

Innovation accelerates responsibly.

Organizations innovate with foresight rather than reactive correction.

AI strengthens civilization instead of destabilizing it.

The Constitution as Living Governance

The defining characteristic of the ZHA Civilization Constitution is adaptability.

Unlike static legal frameworks, it evolves continuously through learning.

The Constitution operates as:

- a guiding philosophy,
- an institutional operating system,
- a societal coordination mechanism,
- a civilization learning framework.

It remains alive because civilization itself evolves.

Civilization Transformation Cycle

Under ZHA implementation:

- Education shapes mindset.
- Mindset shapes innovation.
- Innovation shapes governance.
- Governance protects civilization.

The cycle reinforces itself across generations.

The ZHA Insight

A constitution succeeds not when it controls behavior, but when it inspires responsible action.

The ZHA Civilization Constitution transforms governance from rule enforcement into collective stewardship.

It proves that:

Sustainable AI with Human-Centric Purpose is not an abstract ideal, it is an operational model for civilization itself.

16.12 The ZHA Insight - Governing Intelligence Is Humanity's Next Evolution

The Next Step in Human Civilization

Human civilization has progressed through stages defined by what humanity learned to govern.

In early history, survival depended on governing nature.
Later, stability depended on governing society.
Modern civilization matured through governing institutions and economies.

Today humanity stands at an unprecedented threshold.

For the first time in history, humanity must learn to govern intelligence itself.

Artificial Intelligence marks a transition unlike any previous technological shift.

It does not merely increase human capability.
It multiplies decision-making power.

AI influences:

- how knowledge is created,
- how decisions are executed,
- how economies operate,
- how societies organize,
- how future generations experience reality.

The central challenge of the AI era therefore becomes clear:

Humanity must evolve from creators of intelligence to stewards of intelligence.

Humanity's Governance Evolution

Civilization history can be understood as a sequence of governance maturity.

Each transition required new institutions, ethics, and cultural awareness.

Artificial Intelligence represents the next evolutionary step.

Humanity must now answer a deeper question:

- Who guides intelligence when intelligence begins guiding humanity?
- Why Intelligence Requires Governance

Unlike previous technologies, AI possesses three unique characteristics:

1. Intelligence Scales Instantly
An algorithm deployed once may influence billions of people simultaneously.

A single design decision can reshape:

- markets,
- elections,
- education,
- healthcare outcomes.

Uncoordinated intelligence becomes systemic risk.

2. Intelligence Learns Continuously
AI systems evolve after deployment.

They adapt through:

- data feedback,
- behavioral interaction,
- autonomous optimization.

Traditional governance based on static regulation becomes insufficient.

Governance must become adaptive and anticipatory.

3. Intelligence Shapes Human Behavior
AI does not only execute decisions — it influences human thought itself.

Recommendation systems shape attention.
Automation shapes work identity.
Generative systems shape creativity and knowledge.

When intelligence influences consciousness, governance becomes civilizational.

The Limits of Traditional Governance
Existing governance models were designed for earlier eras.

Regulatory Governance
Responds after harm occurs.

Market Governance
Optimizes efficiency and profit.

Institutional Governance
Manages sectors independently.

None alone can guide systems capable of reshaping humanity's collective future.

The AI era exposes a governance gap:

Humanity possesses powerful intelligence without unified civilizational guidance.

The ZHA Civilization Charter and Constitution as Evolutionary Response

The ZHA Civilization Charter and Constitution emerge as humanity's maturation response.

They introduce a new principle:

Intelligence must evolve within ethical, sustainable, and human-centric boundaries.

These frameworks do not replace national laws or institutions.

Instead, they provide a meta-governance layer guiding how intelligence integrates into civilization.

They ensure:

- intelligence evolves responsibly,
- innovation remains sustainable,
- governance anticipates consequences,
- civilization advances consciously.

From Governing Systems to Governing Intelligence

Traditional governance asked:

- How do we control resources?
- How do we manage populations?
- How do we regulate economies?

ZHA Civilization Governance asks a new question:

How do we guide intelligence so it strengthens humanity rather than overwhelms it?

This transition marks humanity's psychological and ethical evolution.

Governance moves from authority to stewardship.

Responsible Intelligence as Civilizational Maturity

A mature civilization is not defined by technological power alone.

It is defined by how responsibly that power is guided.

The ZHA framework recognizes that:

Intelligence without ethics creates instability.

Innovation without sustainability creates collapse.

Automation without meaning creates societal fragmentation.

Responsible intelligence integrates:

- human dignity,
- planetary stewardship,
- shared prosperity,
- long-term foresight.

Civilization maturity therefore becomes measured by wisdom rather than capability.

Sustainable AI with Human-Centric Purpose

The ZHA insight reframes AI transformation fundamentally.

AI is not simply an engineering challenge.

It is a civilizational responsibility.

Sustainable AI with Human-Centric Purpose ensures that:

- technology amplifies human wellbeing,
- intelligence protects ecological balance,
- innovation benefits future generations,
- progress remains aligned with humanity's deepest values.

AI becomes an instrument of conscious evolution.

Civilization Evolution, Not Governance Reform

The emergence of the ZHA Civilization Charter and Constitution represents more than policy innovation.

It signals humanity entering a new evolutionary phase.

Previous governance models organized society.

ZHA governance organizes intelligence itself.

This transformation shifts civilization:

- from reaction → foresight,
- from competition → collaboration,
- from exploitation → stewardship,

- from technological acceleration → conscious advancement.

Governance becomes an expression of collective wisdom.

The Evolutionary Transition

Humanity's journey can now be understood as:

Creation → Control → Coordination → Conscious Stewardship

The ZHA Civilization model represents the stage of conscious stewardship.

For the first time, humanity deliberately designs systems ensuring intelligence aligns with civilization survival.

The ZHA Insight

Humanity's greatest achievement will not be building intelligent machines.

It will be learning how to live wisely with intelligence.

The future will belong not to the most technologically advanced civilization, but to the civilization capable of governing intelligence responsibly.

The ZHA Civilization Charter and Constitution represent this evolutionary step.

They transform governance into guidance, power into responsibility, and intelligence into a force for collective flourishing.

Chapter Reflection

Civilizations endure when values become systems.

The ZHA Civilization Charter defines humanity's shared purpose.

The ZHA Civilization Constitution ensures that purpose becomes operational reality.

Together they transform Sustainable AI with Human-Centric Purpose from philosophy into global civilization governance.

Artificial Intelligence will shape humanity's future.

The Charter and Constitution ensure humanity remains wise enough to guide it.

CHAPTER XVII

Chapter 17: ZHA World Council and Civilization Index

Introduces institutional structures designed to scale Sustainable AI globally through governance, research, and collaborative leadership.

17.0 Introduction - From Vision to Global Coordination

Throughout this book, the ZHA framework has progressively evolved:

- from mindset transformation,
- to responsible innovation,
- to Sustainable AI governance,
- to civilization charter and constitutional alignment.

Yet a civilization framework becomes meaningful only when supported by living institutions capable of guiding humanity continuously.

Civilizations do not sustain themselves through ideas alone.

They endure through:

- coordinated leadership,
- shared measurement systems,
- collaborative decision-making,
- long-term stewardship structures.

The AI era therefore requires two complementary mechanisms:
A global leadership body capable of guiding intelligence responsibly.

A civilization measurement system capable of evaluating humanity's real progress.

The ZHA World Council and Civilization Index together fulfill these roles.

They transform Sustainable AI from philosophy into global operating reality.

17.1 Why Humanity Needs a World Council in the AI Era

From Fragmented Intelligence to Coordinated Civilization Stewardship

Human civilization has entered a moment unlike any previous era.

Artificial Intelligence no longer belongs to a single nation, organization, or industry.

It operates across borders, cultures, and systems simultaneously.

AI has become global infrastructure.

Artificial Intelligence Beyond Borders

Unlike earlier technologies that spread gradually, AI systems achieve planetary reach immediately upon deployment.

A single algorithm released today can influence billions of people within hours.

AI systems now:

- influence global markets instantly through algorithmic trading,
- shape political discourse via recommendation engines,
- redefine labor structures through automation,
- affect environmental outcomes through energy-intensive computation,
- guide healthcare, education, transportation, and governance decisions.

Intelligence itself has become networked at planetary scale.

Humanity is no longer managing isolated technologies.

Humanity is living inside an interconnected intelligence ecosystem.

The Governance Paradox of the AI Age

Despite AI's global reach, governance structures remain local and fragmented.

Today:

- nations regulate independently, often based on national priorities rather than shared global responsibility,
- corporations innovate competitively, racing for technological advantage,
- academic institutions research in silos, disconnected from deployment realities,
- societies experience uneven technological impact, creating inequality between regions.

This creates a profound imbalance.

Technology has become global.

Governance remains fragmented.

The Emerging Risk: Global Intelligence Without Global Coordination

When intelligence evolves faster than coordination mechanisms, civilization faces systemic risks.

Without coordination:

- ethical standards vary widely,
- safety practices become inconsistent,
- economic disruption spreads unevenly,
- environmental impacts remain unmanaged,
- public trust declines.

This condition may be described as:

Global intelligence without global stewardship.

The risk is not intelligence itself.

The risk is intelligence operating without shared responsibility.

Lessons from History - Power Requires Cooperation

Human history repeatedly demonstrates that powerful systems require cooperative governance.

International Aviation

Air travel became safe only after nations agreed on shared standards:

- air traffic coordination,
- safety protocols,
- global regulatory alignment.

Without cooperation, global aviation would be impossible.

Global Financial Systems

Modern economies depend on coordinated financial institutions:

- international monetary cooperation,
- banking regulations,
- shared economic stabilization mechanisms.

Financial systems require trust sustained through collaboration.

Climate Agreements

Environmental protection demanded global coordination because climate systems ignore political borders.

Nations learned that planetary challenges require collective action.

Internet Governance

The internet succeeded because shared technical standards allowed global interoperability.

No single country owns the internet.
Humanity collectively sustains it.

Artificial Intelligence - A More Powerful System Than All Before

Artificial Intelligence surpasses previous infrastructures because it governs decision-making itself.

AI influences:

- what humans see,
- what humans believe,
- how economies behave,
- how policies evolve,
- how societies organize knowledge.

Unlike aviation or finance, AI shapes human cognition.

Therefore, AI governance cannot remain fragmented.

It requires a coordination model equal to its power.

Why Existing Institutions Are Not Enough

Current global structures were not designed for intelligent systems.

National Governments

Operate within borders.

AI operates globally.

Corporations

Optimize innovation and growth.

Civilization requires sustainability and responsibility.

Academic Institutions

Generate knowledge.

But knowledge alone cannot guide global deployment.

Regulatory Bodies

React after risks emerge.

AI risks evolve continuously.

No single institution can independently guide the Intelligent Age.

Humanity faces an institutional gap.

The Need for a World Council

The AI era demands a new type of global institution.

Not a centralized authority.

Not a political superstructure.

But a civilization coordination platform capable of aligning humanity's relationship with intelligence.

A World Council becomes necessary to:

- create shared ethical direction,
- coordinate Sustainable AI governance,
- foster collaboration across nations and industries,
- anticipate long-term civilization risks,
- maintain public trust in intelligent systems.

This is not control over innovation.

It is stewardship of intelligence.

The ZHA World Council - Humanity's Coordinating Intelligence

The ZHA World Council emerges as a response to this historic necessity.

It represents a new institutional philosophy:

Guidance instead of dominance.
Collaboration instead of competition.
Stewardship instead of regulation alone.

The Council functions as:

- a global foresight body,
- a sustainability governance platform,
- a collaborative leadership ecosystem,
- a civilization intelligence coordinator.

Its purpose is simple yet profound:

To ensure Artificial Intelligence evolves in service of humanity, the planet, and future generations.

Sustainable AI with Human-Centric Purpose

Under ZHA alignment, AI transformation moves beyond technological adoption.

It becomes civilization stewardship.

The World Council ensures that:

- innovation respects human dignity,
- intelligence protects planetary systems,
- economic progress remains inclusive,
- technological evolution strengthens civilization stability.

Sustainable AI becomes humanity's shared project.

The ZHA Insight
Every technological era required new institutions.

The AI era requires humanity's first institution dedicated to governing intelligence itself.

Without coordination, intelligence fragments civilization.

With coordination, intelligence becomes civilization's greatest stabilizing force.

The ZHA World Council represents humanity's transition from competing innovators to collective stewards of intelligence.

17.2 The ZHA World Council - Purpose and Vision

A Civilization Stewardship Institution for the Intelligent Age

The emergence of Artificial Intelligence introduces a profound shift in human history.

For the first time, humanity is not merely governing tools, systems, or economies.

Humanity is shaping intelligence itself.

Such a transformation cannot be guided solely by governments, corporations, or technological communities acting independently.

It requires a new type of institution.

The ZHA World Council is created to fulfill this role.

Beyond Political Authority

The ZHA World Council is not designed as:

- a supranational government,
- a regulatory enforcement body,
- a political alliance,
- or a centralized power structure.

Political institutions govern territory.

The ZHA World Council guides civilization direction.
Its authority does not arise from control.

It arises from collective legitimacy, shared wisdom, and civilizational responsibility.

The Council exists to align humanity's relationship with intelligent systems — ensuring that technological progress strengthens civilization rather than destabilizes it.

A Civilization Stewardship Institution

The ZHA World Council represents a new institutional category:

Civilization Stewardship Institution

Stewardship differs fundamentally from regulation.

In the AI era, humanity requires stewardship because intelligent systems evolve continuously.

Governance must therefore evolve continuously as well.

The Council becomes humanity's shared mechanism for guiding intelligent evolution responsibly.

Purpose of the ZHA World Council

The central purpose of the ZHA World Council is:

To guide the evolution of Sustainable AI aligned with human-centered values and planetary wellbeing.

This purpose recognizes a foundational reality:

Artificial Intelligence will influence every dimension of civilization.

Therefore, its development must remain aligned with:

- human dignity,
- environmental sustainability,
- social inclusion,
- long-term global stability.

The Council ensures that intelligence grows together with wisdom.

Core Mission

The mission of the ZHA World Council is to ensure that Artificial Intelligence advances:

Ethical Advancement

AI systems must respect human rights, fairness, and transparency.

Example:
Guidelines encouraging algorithmic explainability prevent opaque decision systems affecting livelihoods without accountability.

Sustainable Advancement

AI innovation must operate within planetary limits.

Example:
Promoting energy-efficient AI architectures reduces environmental impact while maintaining innovation capability.

Inclusive Advancement

Technological progress must benefit all communities, not only technologically advanced regions.

Example:
Global initiatives enabling AI education access in developing regions prevent widening digital inequality.

Intergenerational Responsibility

AI decisions must consider long-term consequences beyond immediate economic gain.

Example:
Policies encouraging sustainable automation ensure workforce transitions rather than mass displacement.

Through these principles, intelligence becomes a force for enduring civilization progress.

The Council as a Collective Intelligence Forum

Unlike traditional institutions dominated by single sectors, the ZHA World

Council operates as a Collective Intelligence Forum.

It connects diverse perspectives necessary for guiding intelligent civilization.

Participants Include

Educators

- Shape future generations through sustainability mindset education.
- They ensure ethical awareness emerges before technological deployment.

Scientists and Researchers

- Provide evidence-based insight into technological and societal impacts.
- They anticipate long-term risks and opportunities.

Industry Leaders

- Translate principles into real-world innovation and responsible product development.
- They operationalize Sustainable AI practices.

Policymakers

- Align governance frameworks with civilization-scale responsibility.
- They enable supportive regulatory ecosystems rather than restrictive barriers.

Sustainability Practitioners

- Bridge environmental, social, and governance expertise with technological advancement.

- They ensure ESG principles remain embedded within innovation.

Civil Society Representatives

- Bring community perspectives, public concerns, and societal values into decision-making.
- They ensure intelligence serves humanity collectively.

Shared Responsibility Toward Intelligent Systems

The ZHA World Council embodies a new realization:

Artificial Intelligence is no longer owned by creators alone.

Its consequences are shared by humanity.

Therefore responsibility must also be shared.

The Council institutionalizes this principle:

Intelligence is a common civilizational asset requiring collective stewardship.

From Competition to Cooperative Guidance

Traditional technological development has been driven by competition.

While competition accelerates innovation, it often neglects coordination.

The ZHA World Council introduces cooperative guidance.

Organizations continue innovating independently, yet align around shared civilization principles.

Innovation remains dynamic.

Civilization remains stable.

Real-World Scenario - Council in Action

Consider global deployment of advanced generative AI systems.

Without coordinated guidance:

- ethical standards vary,
- misinformation risks increase,
- societal trust declines.

With ZHA World Council collaboration:

- shared ethical deployment frameworks emerge,
- sustainability metrics guide adoption,
- cross-sector learning accelerates,
- public confidence strengthens.

The Council does not restrict innovation.

It enables responsible innovation at global scale.

Vision of the ZHA World Council

The long-term vision of the Council is ambitious yet necessary:

To cultivate a world where:

- intelligence enhances human flourishing,
- innovation protects ecological balance,
- technology strengthens democratic participation,
- global cooperation replaces technological fragmentation.

The Council becomes humanity's conscience in the age of intelligent systems.

The ZHA Insight

Every civilization creates institutions reflecting its level of maturity.

Industrial civilization built economic institutions.
Digital civilization built connectivity institutions.

The Intelligent Age requires stewardship institutions.

The ZHA World Council represents humanity's transition from managing technology to guiding intelligence responsibly.

It symbolizes a profound shift:

Humanity moves from being users of intelligence to custodians of intelligent civilization.

17.3 From Leadership to Stewardship

The Governance Philosophy of the Intelligent Civilization

The Limits of Traditional Leadership

Throughout history, leadership has largely been defined by authority and control.

Traditional leadership structures evolved in environments where:

- resources were scarce,
- power determined stability,
- decisions were centralized,
- outcomes were short-term and measurable.

Leaders governed through hierarchy.

Success was often measured by:

- dominance,
- expansion,
- efficiency,
- competitive advantage.

This model proved effective in earlier eras:

- monarchs governed territory,
- industrial leaders controlled production,
- corporate executives directed markets,
- political systems managed national interests.

However, Artificial Intelligence introduces a radically different reality.

AI systems operate across borders, sectors, and generations simultaneously.

No single leader, organization, or nation can control intelligent systems alone.

The Intelligent Age therefore exposes a fundamental truth:

Control-based leadership cannot guide globally distributed intelligence.

A new leadership paradigm becomes necessary.

The Emergence of Stewardship Leadership

The ZHA World Council introduces a transformational shift:

Leadership evolves into stewardship.

Stewardship does not seek ownership over the future.

It seeks responsibility for the future.

Where leadership traditionally asks:

- How do we win?

Stewardship asks:

- How do we sustain civilization?

Leadership vs Stewardship

Stewardship recognizes that intelligence is now a shared civilizational asset.

Guiding it requires wisdom, collaboration, and foresight.

Why Stewardship Is Essential in the AI Era

Artificial Intelligence amplifies human decisions at unprecedented scale.

A design choice made today can influence billions tomorrow.

Examples include:

- algorithmic recommendations shaping democratic discourse,
- autonomous systems affecting employment structures,
- AI models influencing education and knowledge formation,
- intelligent automation transforming economic stability.

When decisions carry civilizational consequences, leadership must transcend personal or institutional ambition.

Stewardship becomes an ethical necessity.

The Role of ZHA World Council Members

Members of the ZHA World Council do not function as rulers or regulators.

They function as civilization stewards.

Their responsibility is to guide intelligence toward sustainable human progress.

1. Guardians of Long-Term Civilization Wellbeing
Council members evaluate decisions not only for present benefit but for future generations.

They consider:

- environmental sustainability,
- societal resilience,
- intergenerational equity,
- human dignity.

Example:

Before endorsing widespread deployment of autonomous decision systems, stewards examine:

- societal trust implications,
- employment transitions,
- psychological impacts on communities.

Innovation proceeds with foresight rather than urgency alone.

2. Facilitators of Global Collaboration

In the AI era, collaboration becomes survival strategy.

Stewards build bridges between:

- nations,

- industries,
- academic institutions,
- civil society.

They reduce fragmentation by encouraging shared standards and cooperative innovation.

Example:

Instead of competing proprietary climate AI systems, Council stewardship promotes shared research frameworks accelerating planetary sustainability solutions.

Collaboration replaces technological rivalry.

3. Foresight Thinkers Anticipating Technological Consequences

Traditional leadership reacts to change.

Stewardship anticipates it.

Council members engage in long-term foresight:

- assessing emerging AI risks,
- modeling societal impacts,
- anticipating ethical dilemmas,
- preparing humanity for technological transitions.

Example:

Before widespread adoption of AI-generated media, stewards analyze potential misinformation risks and encourage proactive transparency standards.

Civilization becomes prepared rather than surprised.

4. Ethical Navigators Guiding Innovation Direction

Innovation alone does not guarantee progress.

Ethical navigation ensures innovation aligns with human values.

Stewards ask critical questions:

- Does this technology enhance human wellbeing?
- Does it preserve planetary balance?
- Does it strengthen social trust?
- Does it benefit future generations?

Ethics becomes embedded into innovation pathways rather than applied afterward.

Example:

An AI healthcare platform is evaluated not only for accuracy but also for accessibility across socioeconomic groups.

Technology becomes inclusive by design.

Authority Transformed: From Power to Responsibility

The ZHA World Council redefines authority itself.

Authority no longer derives from control over people or resources.

It derives from commitment to collective wellbeing.

This transformation reflects a deeper civilizational maturity:

Power governs systems.
Responsibility sustains civilization.

Council members hold influence not to dominate outcomes but to protect humanity's shared future.

Stewardship as Civilizational Consciousness

Stewardship leadership reflects an evolution in human consciousness.

Humanity moves from:

- ownership → guardianship,
- competition → cooperation,
- extraction → sustainability,
- short-term success → long-term flourishing.

Leaders become caretakers of civilization continuity.

Real-World Scenario - Stewardship Leadership in Action

Imagine global deployment of advanced autonomous AI governance systems.

Traditional leadership might prioritize:

- rapid deployment,
- economic advantage,
- technological prestige.

Stewardship leadership instead evaluates:

- societal readiness,
- ethical safeguards,
- environmental impact,
- democratic transparency.

Deployment proceeds responsibly, maintaining public trust while enabling innovation.

The outcome is sustainable progress rather than technological disruption.

The ZHA Insight

Every civilization evolves when leadership evolves.

The Industrial Age required managers.
The Digital Age required innovators.
The Intelligent Age requires stewards.

The future will belong not to those who control intelligence, but to those who guide it wisely.

The ZHA World Council embodies this transformation.

It represents humanity's shift from leading systems to stewarding civilization itself.

17.4 Structure of the ZHA World Council

Designing a Distributed Governance Architecture for Sustainable AI Civilization

From Centralized Authority to Distributed Global Representation

Traditional governance institutions were designed for earlier civilizations where power, resources, and decision-making were geographically centralized.

Artificial Intelligence fundamentally changes this reality.

Intelligence today is:

- distributed across cloud networks,
- developed by global teams,
- deployed simultaneously across continents,
- influencing societies beyond national jurisdiction.

A centralized governance model cannot effectively guide a decentralized intelligence ecosystem.

Therefore, the ZHA World Council adopts a distributed representation structure.

Rather than concentrating authority, the Council distributes stewardship across specialized global bodies — each responsible for guiding a critical dimension of Sustainable AI civilization.

This structure reflects a key ZHA principle:

Civilization governance must mirror the interconnected nature of intelligence itself.

Each council acts independently yet collaboratively, forming an integrated global stewardship system.

1. Civilization Stewardship Board

Strategic Heart of the ZHA World Council

The Civilization Stewardship Board functions as the Council's highest foresight and ethical direction body.

Its role is not operational management but civilizational guidance.

Core Purpose
To ensure AI evolution aligns with humanity's long-term survival, dignity, and flourishing.

Key Responsibilities

Defining Global Ethical Principles

The Board develops foundational ethical frameworks guiding AI development worldwide.

These principles address:

- human rights in algorithmic decision-making,
- responsible data use,
- technological dignity,
- planetary responsibility.

Example:
Establishing global guidance that autonomous systems must remain human-overrideable to protect accountability.

Civilization Foresight Planning

The Board anticipates long-term technological consequences.

It studies emerging risks such as:

- AI-driven economic inequality,
- cognitive manipulation,
- environmental computing costs,
- human identity transformation.

Foresight shifts civilization from reactive governance to anticipatory stewardship.

Aligning International Cooperation

The Board encourages collaboration between nations, institutions, and industries.

Rather than imposing standards, it facilitates shared alignment.

Example:
Multi-country agreements on Sustainable AI energy standards preventing environmental damage from large-scale model training.

Civilizational Role

The Stewardship Board acts as humanity's strategic conscience in the Intelligent Age.

2. Sustainable AI Governance Forum

Operational Governance Engine

While the Stewardship Board defines direction, the Sustainable AI Governance Forum translates vision into practical governance frameworks.

Purpose

To design governance models enabling responsible AI adoption across organizations and nations.

Core Responsibilities

ESG-Aligned AI Policy Development

The Forum develops guidance integrating Environmental, Social, and Governance principles into AI transformation.

Policies address:

- sustainable infrastructure usage,
- workforce transition ethics,
- inclusive technology access.

Ethical Deployment Frameworks

Ensures AI systems are deployed responsibly.

Frameworks include:

- transparency standards,
- bias monitoring practices,
- human oversight mechanisms.

Example:
Healthcare AI deployment frameworks requiring explainability before clinical usage.

Risk Anticipation Models

Instead of regulating after crises, the Forum identifies risks early through scenario modeling.

Examples:

- misinformation escalation risks,
- autonomous weapon governance,
- systemic algorithmic bias.

Governance evolves alongside innovation.

Civilizational Role

The Forum becomes the global operating system for Sustainable AI governance.

3. Research and Knowledge Council

Civilization Intelligence Network

Knowledge creation is essential for guiding intelligent civilization.

The Research and Knowledge Council coordinates global intellectual collaboration.

Purpose

To transform research into civilization intelligence rather than isolated academic output.

Collaborative Participants

- universities,
- research laboratories,
- think tanks,
- innovation hubs,
- sustainability institutes.

Strategic Focus Areas

Climate Intelligence

AI models predicting environmental risk and supporting planetary restoration.

Example:
Global AI climate modeling predicting drought patterns and guiding sustainable agriculture policies.

Human-AI Collaboration

Research exploring how humans and intelligent systems work together ethically and effectively.

Focus:

- augmented decision-making,
- cognitive wellbeing,
- human creativity enhancement.

Sustainable Infrastructure

Developing energy-efficient computing systems and environmentally responsible digital infrastructure.

Social Impact Modeling

Studying how AI affects:

- employment,
- education,
- social cohesion,
- democratic trust.

Civilizational Role

The Council becomes humanity's collective brain, ensuring knowledge serves civilization survival.

4. Education and Human Development Council

Building the Sustainability Mindset Generation

Technological governance begins with human mindset.

The Education and Human Development Council ensures that future generations develop responsibility alongside capability.

Purpose

To embed sustainability mindset education globally.

Key Responsibilities

Curriculum Alignment

Integrates Sustainable AI principles into global education systems.

Students learn:

- ethical innovation,
- environmental responsibility,
- collaborative problem solving,
- civilization stewardship.

Leadership Training

Develops leaders capable of guiding intelligent systems responsibly.

Programs cultivate:

- systems thinking,
- ethical decision-making,
- sustainability leadership.

Certification Ecosystems - ZSPC

Through ZSPC certifications, individuals become recognized Sustainability Mindset Practitioners.

Outcome:
A globally aligned workforce prepared for responsible AI transformation.

Civilizational Role

Education becomes civilization's long-term stability mechanism.

5. Industry and Innovation Council

Aligning Innovation with Civilization Responsibility

Industry drives technological acceleration.

The Industry and Innovation Council ensures that innovation advances sustainably.

Purpose

To transform corporations from technology competitors into civilization partners.

Core Functions

Promoting Sustainable AI SDLC Standards

Organizations adopt development lifecycles integrating:

- ESG evaluation,
- ethical design checkpoints,
- sustainability impact assessment.

Example:
Product teams evaluating environmental cost before scaling AI infrastructure.

Sharing ESG Best Practices

Corporations collaborate rather than reinvent responsible innovation independently.

Shared learning accelerates global adoption.

Enabling Cooperative Innovation Platforms

Companies collaborate on civilization-scale challenges such as:

- climate resilience,
- healthcare accessibility,
- sustainable cities.

Innovation shifts from competition toward collective progress.

Civilizational Role

Industry becomes an engine of sustainable progress rather than unchecked acceleration.

Integrated Council Architecture

Together, the five bodies form an interconnected governance ecosystem:

The structure mirrors the ZHA philosophy:
Mindset → Knowledge → Governance → Innovation → Civilization Stability

The ZHA Insight

Civilizations endure when leadership, knowledge, education, governance, and innovation operate in harmony.

The ZHA World Council does not centralize power.

It distributes responsibility.

It transforms governance from control over systems into stewardship of humanity's intelligent future.

17.5 The Need for a Civilization Index

Measuring Progress in the Age of Intelligent Civilization

Human civilization has always relied on measurement to understand progress.

What societies measure ultimately shapes what they become.

For more than a century, modern civilization has evaluated success primarily through economic indicators, most prominently Gross Domestic Product (GDP).

GDP transformed global governance by providing a simple answer to a complex question:

- How productive is an economy?

It enabled nations to compare growth, guide policy decisions, and track economic expansion.

Yet the Artificial Intelligence era reveals a profound limitation:

Economic growth alone does not measure civilization health.

The Limits of GDP in the Intelligent Age

GDP measures production.

It does not measure purpose.

GDP increases when:

- industries expand,
- consumption rises,
- technological output accelerates.

However, GDP does not capture whether progress improves humanity's long-term wellbeing.

It fails to measure:

Environmental Stability

An economy may grow while ecosystems degrade.

AI-driven industries may expand computing infrastructure while increasing energy consumption and carbon emissions.

GDP records growth.
The planet records strain.

Social Trust

Technological progress may coexist with declining societal cohesion.

AI-powered platforms can amplify polarization, misinformation, and distrust.

Economic productivity rises while social stability weakens.

GDP remains silent.

Ethical Innovation

Innovation speed often becomes the dominant metric.

Yet technologies deployed without ethical safeguards may produce unintended harm.

GDP rewards innovation volume, not innovation responsibility.

Long-Term Wellbeing

Automation may increase productivity while reducing meaningful employment or human fulfillment.

Economic indicators show efficiency gains.

Human wellbeing may decline.

The Civilization Measurement Gap

Humanity now faces a measurement crisis.

The question is no longer:

- How fast are we growing?

The AI era demands a deeper question:

- Are we progressing wisely, or merely progressing rapidly?

Artificial Intelligence accelerates civilization faster than any technology in history.

Without new measurement systems, humanity risks optimizing for speed while neglecting sustainability.

Civilization may advance technologically while regressing socially and environmentally.

Why the AI Era Requires a New Metric

Artificial Intelligence transforms civilization across multiple dimensions simultaneously:

- economic systems,
- governance structures,
- environmental impact,
- cultural behavior,
- human cognition itself.

Traditional indicators designed for industrial economies cannot capture these multidimensional effects.

AI introduces outcomes that GDP cannot quantify:

- algorithmic fairness,
- digital inclusion,
- ethical governance,
- cognitive wellbeing,

- sustainability of intelligent systems.

A civilization shaped by intelligence requires a civilization-scale measurement framework.

From Economic Growth to Civilization Health

The ZHA framework proposes a fundamental shift:

Progress must be measured not only by wealth creation, but by civilization wellbeing.

A healthy civilization demonstrates balance across:

- prosperity,
- sustainability,
- inclusivity,
- trust,
- responsibility,
- long-term resilience.

Economic success becomes one dimension of a broader human story.

The ZHA Civilization Index

The ZHA Civilization Index emerges as humanity's response to the Intelligent Age.

It answers a critical need:

To evaluate whether intelligence strengthens civilization rather than destabilizes it.

The Index measures progress across integrated civilizational dimensions:

- Environmental sustainability,

- Social inclusion and trust,
- Governance transparency,
- Ethical innovation,
- Human wellbeing.

Instead of asking how much humanity produces, the Index asks:

How responsibly humanity evolves.

A Shift in Civilizational Thinking

The introduction of a Civilization Index represents a philosophical transformation.

Industrial civilization measured productivity.
Digital civilization measured connectivity.
Intelligent civilization must measure responsibility.

Measurement becomes guidance.

Nations, organizations, and institutions gain visibility into the long-term consequences of technological decisions.

Progress becomes intentional rather than accidental.

Real-World Illustration

Consider two countries with identical GDP growth.

Country A:

- deploys energy-intensive AI without sustainability safeguards,
- experiences workforce displacement,
- sees rising social distrust.

Country B:

- invests in Sustainable AI,
- reskills workers,
- ensures algorithm transparency,
- reduces environmental impact.

GDP views both as equally successful.

The Civilization Index reveals which nation is building a sustainable future.

Civilization Intelligence Through Measurement

The ZHA Civilization Index functions as humanity's collective feedback system.

It allows civilization to:

- learn from its actions,
- detect emerging risks,
- align innovation with values,
- course-correct before crises emerge.

Measurement evolves from accounting performance to guiding civilization wisdom.

The ZHA Insight

Civilizations become what they measure.

If humanity measures only economic output, it optimizes for consumption.

If humanity measures technological dominance, it optimizes for competition.

If humanity measures sustainability, wellbeing, and responsibility, it evolves toward conscious civilization.

The ZHA Civilization Index ensures that humanity measures what truly sustains life, society, and future generations.

17.6 The ZHA Civilization Index - Measuring Civilization Health

A Holistic Framework for Evaluating Progress in the Age of Sustainable AI

From Economic Measurement to Civilization Measurement

Human civilization has long relied on numerical indicators to evaluate progress.

Industrial societies measured production.
Digital societies measured connectivity.
Financial systems measured growth.

However, the emergence of Artificial Intelligence introduces a deeper challenge:

Technology now shapes civilization itself.

AI influences how humans work, learn, interact, govern, and even perceive reality.

Therefore, progress can no longer be measured by isolated metrics.

Civilization requires a holistic measurement system, one capable of evaluating whether intelligence strengthens humanity, society, and the planet simultaneously.

The ZHA Civilization Index fulfils this role.

It evaluates civilization not only by what humanity builds, but by what humanity becomes through intelligence.

Purpose of the Civilization Index

The ZHA Civilization Index answers a foundational question:

Is intelligence advancing civilization health or weakening it?

Rather than measuring isolated success, the Index examines the balance between:

- technological advancement,
- environmental sustainability,
- social equity,
- governance responsibility,
- human wellbeing.

Civilization health emerges only when all dimensions evolve together.

Core Dimensions of the ZHA Civilization Index

Environmental Sustainability

Measuring Planetary Stewardship

Artificial Intelligence consumes physical resources:

- computational energy,
- data center infrastructure,
- rare materials,
- global supply chains.

Technological intelligence must therefore operate within planetary limits.

The Environmental Sustainability dimension evaluates whether AI development protects ecological systems rather than exhausting them.

Key Indicators

- carbon efficiency of AI models,
- renewable energy usage in data centers,

- biodiversity protection initiatives,
- sustainable resource utilization,
- circular technology practices,
- energy-efficient digital infrastructure.

Example:
An organization transitions from large energy-intensive models to optimized architectures requiring significantly lower computation.

Result:

- reduced carbon footprint,
- improved environmental score,
- sustainable innovation leadership.

Innovation becomes environmentally intelligent.

Civilizational Insight

Progress that damages the planet cannot sustain civilization.

Environmental stewardship becomes a prerequisite for intelligent advancement.

Social Inclusion

Measuring Equity in the Intelligent Society

AI transformation reshapes employment, education, and opportunity distribution.

Without inclusion, intelligence amplifies inequality.

The Social Inclusion dimension evaluates whether technological progress benefits society broadly rather than concentrating advantage.

Key Indicators:

- access to AI education and literacy,
- workforce transition and reskilling programs,
- gender equality in technological participation,
- accessibility adoption,
- equitable technology deployment,
- community participation in governance.

Example:
A company introducing automation invests in workforce reskilling programs enabling employees to transition into AI-supported roles.

Outcome:

- reduced unemployment risk,
- increased workforce resilience,
- higher inclusion score.

Technology becomes socially enabling rather than disruptive.

Civilizational Insight

A civilization succeeds when intelligence expands opportunity rather than exclusion.

Governance Transparency

Measuring Responsible Leadership

AI systems increasingly influence decisions affecting individuals and societies.

Opaque intelligence undermines trust.

The Governance Transparency dimension evaluates whether leadership ensures accountability and openness in intelligent systems.

Key Indicators:

- algorithm transparency standards,
- ethical audit compliance,
- explainability mechanisms,
- public trust indicators,
- regulatory cooperation,
- accountability frameworks.

Example:

A public-sector AI decision system undergoes independent ethical auditing and publishes explainability reports accessible to citizens.

Outcome:

- increased societal trust,
- strengthened democratic legitimacy,
- improved governance rating.

Governance evolves from authority to accountability.

Civilizational Insight

Trust becomes the foundation of intelligent civilization.

Ethical Innovation

Measuring Responsible Technological Creation

Innovation has historically prioritized speed and competitiveness.

The AI era requires innovation guided by responsibility.

The Ethical Innovation dimension evaluates how technology is designed — not merely how quickly it is deployed.

Key Indicators:

- human-centered product design,
- bias mitigation practices,
- sustainability-first engineering,
- ethical risk assessments,
- inclusive innovation methodologies,
- long-term societal impact evaluation.

Example:
An AI healthcare system is designed to minimize bias across demographic groups before deployment.

Result:

- equitable outcomes,
- ethical reliability,
- improved innovation score.

Innovation becomes a tool for justice rather than advantage.

Civilizational Insight

The quality of innovation matters more than the speed of innovation.

Human Wellbeing

Measuring the Ultimate Outcome of Civilization

Technology exists to enhance human life.

If intelligence advances while wellbeing declines, civilization fails its purpose.

The Human Wellbeing dimension evaluates the lived human experience resulting from technological transformation.

Key Indicators:

- mental health outcomes,
- quality and meaning of work,
- education accessibility,
- digital wellbeing,
- community resilience,
- societal happiness indicators.

Example:
Organizations implementing AI-assisted workflows reduce employee burnout while increasing creativity and learning opportunities.

Outcome:

- improved wellbeing,
- sustainable productivity,
- healthier organizational culture.

Technology supports human flourishing.

Civilizational Insight

The ultimate measure of progress is not efficiency, it is human wellbeing.

Integrated Civilization Health Assessment

The ZHA Civilization Index operates holistically.

No single dimension determines success.

True civilization health emerges when:

- environmental systems remain stable,
- societies remain inclusive,
- governance remains trustworthy,
- innovation remains ethical,
- humans remain fulfilled.

Balanced progress replaces fragmented advancement.

How the Index Guides Decision-Making

The Civilization Index enables:

Governments
align national AI strategies with sustainability goals.

Organizations
evaluate innovation impact beyond financial performance.

Educational Institutions
measure societal readiness for intelligent transformation.

Global Institutions
identify emerging civilization risks early.

Measurement becomes a tool for wisdom.

From Metrics to Civilization Intelligence

The ZHA Civilization Index transforms measurement itself.

It functions as humanity's collective self-awareness system.

Civilization learns from its outcomes.

Intelligence becomes guided evolution rather than uncontrolled acceleration.

The ZHA Insight

Civilizations decline when they measure growth without meaning.

Civilizations flourish when measurement reflects shared human values.

The ZHA Civilization Index ensures humanity measures:

- responsibility,
- sustainability,
- inclusion,
- wellbeing,
- conscious progress.

It converts Sustainable AI with Human-Centric Purpose into measurable reality.

17.7 Civilization Index vs Traditional Metrics

Redefining Success in the Age of Sustainable AI Civilization

The Measurement Problem of Modern Civilization

Every civilization is shaped by the metrics it chooses to prioritize.

Measurement is never neutral.

What societies measure determines:

- what leaders optimize,
- what organizations reward,
- what technologies accelerate,
- and ultimately, what civilization becomes.

For more than a century, modern development models have relied primarily on industrial-era indicators:

- economic growth,
- production efficiency,
- market expansion,
- technological acceleration.

These metrics successfully powered industrialization and globalization.

However, the Artificial Intelligence era reveals a critical limitation:

Traditional metrics measure activity, not wisdom.

As intelligent systems begin shaping society itself, humanity must evolve its definition of progress.

The ZHA Civilization Index represents this evolution.

From Industrial Success to Civilizational Success

The transition introduced by the ZHA Civilization Index can be understood as a shift from economic optimization to civilization optimization.

Each shift reflects a deeper transformation in human priorities.

GDP Growth → Civilization Wellbeing

Traditional View

GDP measures the monetary value of goods and services produced.

It rewards expansion, consumption, and economic output.

Yet GDP increases even when:

- ecosystems degrade,
- social inequality widens,
- mental health declines,
- technological disruption destabilizes communities.

GDP measures economic motion, not societal health.

ZHA Civilization Perspective

The Civilization Index evaluates wellbeing outcomes instead of economic volume alone.

Civilization Wellbeing includes:

- environmental balance,
- social trust,
- equitable opportunity,
- sustainable prosperity,
- human dignity.

Example:

Two countries experience identical economic growth:

One grows through unsustainable resource extraction.

Another grows through renewable innovation and inclusive technology.

GDP treats both equally.

The Civilization Index distinguishes which society is building a sustainable future.

Insight

Economic growth becomes meaningful only when it improves civilization wellbeing.

Productivity → Human Flourishing

Traditional View

Productivity measures how efficiently humans and machines produce output.

Automation historically aimed to maximize efficiency.

However, productivity gains may also produce:

- job displacement,
- burnout,
- loss of purpose,
- widening inequality.

Efficiency alone does not guarantee fulfillment.

ZHA Civilization Perspective

Human Flourishing measures whether technological progress enhances human life.

Indicators include:

- meaningful work,
- lifelong learning opportunities,
- creativity and innovation capacity,
- mental wellbeing,
- work–life harmony.

Example: An organization introduces AI automation.

Traditional success:
Higher output per employee.

Civilization success:
Employees transition into creative, analytical, and socially meaningful roles supported by AI.

Technology augments humanity rather than replacing it.

Insight

The goal of intelligence is not to replace humans but to help humans flourish.

Market Expansion → Planetary Sustainability

Traditional View

Market expansion rewards growth in consumption and production scale.

Global markets expanded rapidly during industrial and digital eras.

Yet unchecked expansion contributed to:

climate instability,

biodiversity loss,

resource depletion.

Economic expansion without ecological awareness threatens civilization survival.

ZHA Civilization Perspective

Planetary Sustainability evaluates whether economic and technological systems operate within Earth's ecological limits.

Measures include:

- carbon-efficient AI infrastructure,
- sustainable supply chains,
- circular economy adoption,
- biodiversity protection.

Example:

Technology companies collaborate to build energy-efficient AI models rather than competing solely for computational scale.

Innovation continues while environmental impact decreases.

Insight

A civilization cannot prosper on a depleted planet.

Innovation Speed → Responsible Intelligence

Traditional View

Innovation success has often been defined by speed:

- fastest deployment,
- first-to-market advantage,
- rapid scaling.

However, rapid innovation without ethical reflection can produce:

- misinformation crises,
- algorithmic bias,
- social polarization,
- unintended societal harm.

Speed without responsibility creates instability.

ZHA Civilization Perspective

Responsible Intelligence evaluates innovation based on its societal impact.

Responsible Intelligence ensures:

- transparency,
- ethical safeguards,
- sustainability integration,
- human-centered outcomes.

Example:
A generative AI platform delays deployment until fairness testing, bias audits, and energy optimization are completed.

Innovation progresses responsibly rather than recklessly.

Insight

The future belongs not to the fastest innovators, but to the most responsible innovators.

The Civilization Shift: From Growth to Wisdom

The ZHA Civilization Index reframes success itself.

Industrial civilization pursued expansion.
Digital civilization pursued connectivity.
Intelligent civilization must pursue wisdom.

Success becomes defined by balance:

- growth aligned with sustainability,
- innovation aligned with ethics,
- intelligence aligned with humanity.

Why This Shift Matters in the AI Era

Artificial Intelligence amplifies whatever civilization values.

If humanity values only:

- profit,
- speed,
- dominance,
- AI accelerates instability.

If humanity values:

- wellbeing,
- sustainability,
- responsibility,
- AI accelerates civilization evolution.

Measurement therefore becomes civilization guidance.

Real-World Illustration

Consider two global technology ecosystems:

Traditional Metric System

- rapid AI deployment,
- maximum market capture,
- rising environmental cost,
- declining public trust.

ZHA Civilization Index System

- sustainable AI infrastructure,
- inclusive innovation,
- transparent governance,
- improved societal wellbeing.

Both advance technologically.

Only one advances civilizationally.

The ZHA Insight

Metrics are silent philosophers.

They shape civilization without announcing their influence.

The ZHA Civilization Index replaces narrow economic measurement with holistic civilizational awareness.

It teaches humanity to ask:

Not How fast are we growing?
but
How wisely are we evolving?

17.8 How the World Council Uses the Civilization Index

From Measurement to Civilizational Guidance

The Role of Measurement in the Intelligent Age

Measurement has always shaped governance.

In earlier eras, governments measured:

- population,
- economic output,
- military strength,
- industrial productivity.

These metrics helped manage nations.

However, Artificial Intelligence transforms civilization into an interconnected global system where decisions made in one region influence humanity everywhere.

The ZHA World Council therefore requires more than statistics.

It requires civilization intelligence, a continuous understanding of how technological progress affects humanity, society, and the planet.

The ZHA Civilization Index becomes the Council's primary instrument for achieving this understanding.

It functions not as a control mechanism, but as a navigation system for civilization.

Monitoring Global Sustainable AI Adoption

One of the World Council's central responsibilities is observing how Artificial Intelligence evolves worldwide.

The Civilization Index allows the Council to monitor:

- adoption of Sustainable AI practices,
- integration of ESG principles into innovation,
- global accessibility of intelligent technologies,
- environmental impact of AI infrastructure,
- societal readiness for intelligent transformation.

Rather than focusing solely on technological capability, the Council evaluates how responsibly intelligence is being deployed.

Example:
If multiple nations rapidly deploy AI automation without workforce transition programs, the Index reveals declining social inclusion indicators.

The Council recognizes early signs of societal imbalance.

Monitoring enables prevention rather than crisis response.

Identifying Emerging Civilization Risks

Artificial Intelligence introduces risks that evolve faster than traditional governance mechanisms.

The Civilization Index functions as an early warning system.

It identifies patterns such as:

- increasing energy consumption from large-scale AI systems,
- widening digital inequality between regions,
- declining public trust in algorithmic decisions,
- ethical risks from autonomous technologies,
- societal stress caused by rapid technological disruption.

Instead of waiting for visible crises, the World Council detects systemic trends.

Example:
Rising misinformation indicators across multiple regions may signal algorithmic amplification risks.

The Council initiates global dialogue before instability escalates.

Civilization governance becomes anticipatory.

Guiding Policy Recommendations

The ZHA World Council does not impose laws.

Its influence lies in providing informed, evidence-based guidance.

Using Civilization Index insights, the Council develops recommendations for:

- national AI strategies,
- ESG-aligned industry standards,
- ethical deployment practices,
- education transformation initiatives,
- sustainability-focused innovation policies.

Governments, organizations, and institutions voluntarily align because recommendations emerge from shared data and collective understanding.

Example:
If environmental metrics reveal high carbon costs from AI training, the Council recommends global standards for energy-efficient model development.

Innovation continues - but responsibly.

Policy evolves through collaboration rather than coercion.

Encouraging Collaborative Improvement

Traditional rankings often create competition.

The ZHA Civilization Index is intentionally designed differently.

Its purpose is collective improvement, not comparison.

The Council uses the Index to:

- share best practices between nations,
- connect organizations facing similar challenges,
- promote cooperative research initiatives,
- accelerate global learning.

Civilization advances when participants learn from one another.

Example:
A country achieving high workforce transition success shares its reskilling model globally through Council coordination.

Other regions adopt and adapt the approach.

Progress spreads organically.

Learning Instead of Punishment

A defining principle of the ZHA Civilization Index is that it does not punish underperformance.

Punitive systems often lead to:

- data manipulation,
- resistance to participation,
- fear-driven compliance.

The ZHA model replaces judgment with learning.

Low scores indicate opportunity, not failure.

The Council's response becomes:

- support,
- collaboration,
- guidance,
- knowledge sharing.

Measurement becomes an invitation to improve rather than a mechanism of control.

Alignment Over Enforcement

The Civilization Index promotes voluntary alignment.

Organizations and nations participate because alignment produces tangible benefits:

- increased public trust,
- sustainable economic growth,
- long-term stability,
- global credibility.

The World Council acts as a facilitator of alignment rather than an enforcer of compliance.

This approach reflects the ZHA philosophy:

Civilization evolves through shared purpose, not imposed authority.

Measurement as Civilization Navigation

The Civilization Index functions similarly to a global navigation system.

It answers questions such as:

- Where is humanity progressing sustainably?
- Where are risks emerging?
- Which innovations strengthen civilization?
- Which practices require reconsideration?

Instead of steering individual actors, the Index illuminates the path forward.

Humanity collectively adjusts direction.

Real-World Scenario : The Index in Action

Imagine a global surge in AI adoption across multiple industries.

The Civilization Index detects:

- rising productivity,
- but declining wellbeing indicators,
- increasing environmental energy consumption.

The World Council convenes cross-sector discussions involving:

- industry leaders,

- researchers,
- educators,
- policymakers.

Outcomes include:

- sustainable AI infrastructure standards,
- workforce wellbeing policies,
- collaborative environmental initiatives.

The crisis is avoided before it fully emerges.

This is governance through intelligence rather than reaction.

The Transformation of Governance

Through the Civilization Index, governance itself evolves.

Old Model:
Regulation reacts after harm occurs.

ZHA Model:
Measurement guides responsible evolution continuously.

Governance becomes dynamic, adaptive, and learning-oriented.

Civilization develops self-awareness.

The ZHA Insight

The greatest advancement of intelligent civilization is not building smarter machines.

It is building smarter ways of learning from ourselves.

The ZHA World Council uses the Civilization Index to help humanity observe its own evolution.

Measurement becomes wisdom.

Data becomes conscience.

Intelligence becomes guided progress.

17.9 Distributed Global Participation

Civilization Governance as Collective Intelligence

From Centralized Governance to Participatory Civilization

Traditional governance models were built on centralized authority.

Decisions flowed from:

- governments to citizens,
- executives to employees,
- institutions to society.

This structure functioned effectively when systems were localized and slow-moving.

However, Artificial Intelligence transforms civilization into a continuously evolving global network.

Intelligence today emerges not from a single authority, but from millions of interconnected contributors:

- students learning,
- engineers designing,
- communities interacting,
- organizations deploying,
- societies adapting.

In such an environment, governance cannot remain centralized.

It must become participatory.

The ZHA Civilization Index introduces a new principle:

Civilization governance is strongest when humanity participates collectively in guiding intelligence.

The Civilization Index as a Participatory System

The ZHA Civilization Index is not a passive measurement tool operated by a small group of experts.

It is a living system shaped by global participation.

Every stakeholder affected by Artificial Intelligence becomes a contributor to civilization intelligence.

Participation transforms governance from hierarchical decision-making into shared awareness.

Measurement becomes collaborative reflection.

Contributors to Civilization Intelligence

1. Educational Institutions - Generational Insight

Schools, universities, and learning ecosystems contribute data reflecting the readiness of future generations.

They provide insight into:

- AI literacy levels,
- sustainability mindset adoption,
- ethical awareness among students,
- accessibility of education,
- future workforce preparedness.

Example:
Universities integrating responsible AI curricula report improved student participation in sustainability innovation projects.

The Civilization Index reflects rising civilization readiness.

Education becomes civilization foresight.

2. Corporations - Operational Reality
Organizations deploying AI systems contribute real-world operational insights.

They report:

- ESG performance outcomes,
- workforce transformation results,
- sustainable SDLC practices,
- ethical innovation metrics,
- environmental impact of AI infrastructure.

Example:
A company sharing data on energy-efficient model training helps establish global benchmarks for sustainable computing.

Industry becomes an active steward of civilization progress.

3. Governments - Policy Alignment

Governments contribute governance indicators measuring:

- transparency levels,
- public trust metrics,
- inclusive policy adoption,
- national sustainability strategies,
- regulatory collaboration.

Rather than competing policy models, nations learn from one another through shared visibility.

Example:
A country improving algorithm transparency inspires similar governance reforms globally.

Policy evolves through learning instead of competition.

4. Research Organizations - Knowledge Intelligence

Research institutions contribute evidence-based analysis.

They provide:

- societal impact studies,
- climate modeling insights,
- human–AI collaboration research,
- long-term technological risk assessment.

Research transitions from isolated academic output into civilization guidance.

Example:
Global research collaboration identifies early risks of AI-driven misinformation ecosystems, allowing proactive mitigation.

Knowledge becomes collective foresight.

5. Communities - Lived Human Experience

Perhaps the most transformative contribution comes from communities themselves.

Citizens provide insight into:

- wellbeing outcomes,
- social cohesion,
- digital trust,
- accessibility challenges,
- cultural impacts of technology.

Communities ensure that civilization measurement reflects lived reality rather than institutional abstraction.

Example:
Community feedback revealing digital exclusion among rural populations drives global inclusion initiatives.

Human experience becomes a governance signal.

Collective Intelligence as Governance

When contributions from all stakeholders integrate, governance evolves into collective intelligence.

Civilization becomes capable of learning about itself continuously.

Instead of isolated decision-makers attempting to predict the future, humanity develops shared awareness.

The ZHA World Council synthesizes these contributions into actionable insight.

Governance shifts from authority-based control to intelligence-based coordination.

Distributed Responsibility

Distributed participation produces distributed responsibility.

No single institution owns the future of Artificial Intelligence.

Responsibility becomes shared across:

- educators shaping values,
- innovators designing technology,
- leaders guiding deployment,
- citizens experiencing outcomes.

This distributed stewardship reflects a mature civilization capable of guiding its own evolution.

Technology Enabling Participation

Artificial Intelligence itself enables participatory governance.

AI-supported platforms can:

- aggregate global sustainability data,
- analyze social impact trends,
- visualize civilization health indicators,
- enable transparent public engagement.

Technology becomes both the subject and facilitator of governance.

Real-World Scenario, Participatory Civilization Governance

Imagine a global Sustainable AI ecosystem:

- universities share educational readiness metrics,
- corporations publish ESG innovation data,
- governments provide policy transparency reports,
- research institutes analyze emerging risks,
- communities contribute wellbeing feedback.

The Civilization Index synthesizes these signals.

The ZHA World Council identifies patterns and guides collective action.

Humanity governs intelligence together.

The Cultural Transformation of Governance

Distributed participation produces a profound cultural shift.

Citizens move from passive observers of technological change to active contributors in shaping civilization.

Governance becomes:

- collaborative rather than hierarchical,
- adaptive rather than rigid,
- inclusive rather than exclusive.

Civilization becomes self-aware.

The ZHA Insight

The Intelligent Age demands a new understanding of governance:

No single leader is wise enough to guide civilization alone.

Wisdom emerges collectively.

The ZHA Civilization Index transforms humanity into a learning civilization, one capable of observing its own trajectory and consciously adjusting its future.

Distributed participation converts intelligence into shared human responsibility.

17.10 Real-World Scenario - Civilization Index in Practice

From Measurement to Conscious National Transformation

The Transition from Reactive Development to Intentional Progress

For much of modern history, nations have progressed through trial and error.

Policies were often shaped after crises emerged:

- environmental degradation prompted regulation,
- unemployment triggered welfare reform,
- technological disruption forced reactive legislation,
- social instability led to corrective governance.

Progress occurred, but frequently at significant cost.

The ZHA Civilization Index introduces a new approach:

Civilization evolves through awareness rather than reaction.

A nation aligned with ZHA principles does not wait for disruption.

It continuously evaluates how Artificial Intelligence influences civilization health and adjusts direction proactively.

Progress becomes intentional.

The National Adoption of ZHA Alignment

Imagine a nation choosing to adopt the ZHA Civilization Charter and implement the Civilization Index as part of its national development strategy.

Rather than measuring success only through GDP growth, the nation evaluates five interconnected dimensions of intelligent civilization.

The government establishes a cross-sector Civilization Stewardship Council including:

- educators,
- industry leaders,
- sustainability experts,
- AI researchers,
- community representatives,
- policymakers.

The Civilization Index becomes the nation's strategic compass.

Step 1 - Evaluating AI Energy Consumption Across Industries

Artificial Intelligence infrastructure rapidly expands across sectors:

- finance,
- healthcare,
- manufacturing,
- transportation,
- public administration.

The Civilization Index assesses:

- energy demand of AI models,
- carbon emissions from data centers,
- infrastructure sustainability,
- renewable energy adoption.

Findings:
Analysis reveals that large-scale AI deployments significantly increase national energy consumption.

Action Taken

The nation introduces:

- incentives for energy-efficient AI models,

- renewable-powered data centers,
- sustainable computing standards.

Outcome:
Innovation continues while environmental impact decreases.

Technology growth aligns with planetary stewardship.

Step 2 - Assessing Workforce Impact of Automation

Automation improves productivity but can destabilize employment ecosystems.

The Civilization Index evaluates:

- job displacement patterns,
- emerging skill gaps,
- workforce wellbeing indicators,
- regional inequality risks.

Findings:
Automation disproportionately affects certain industries and geographic regions.

Action Taken

Government and industry collaborate to launch:

- national reskilling initiatives,
- AI-assisted workforce transition programs,
- lifelong learning systems.

Outcome:

Workers transition into higher-value roles rather than experiencing exclusion.

Automation strengthens society instead of fragmenting it.

Step 3 - Measuring Education Access to AI Literacy

Civilization sustainability depends on preparing future generations.

The Index evaluates:

- AI education accessibility,
- digital literacy rates,
- sustainability mindset adoption,
- inclusivity of educational opportunities.

Findings:
Urban populations gain AI skills faster than rural communities.

Action Taken

Educational reform introduces:

- universal AI literacy curriculum,
- sustainability mindset education,
- teacher training programs,
- community learning hubs.

Outcome:
Technological capability becomes inclusive.

Intelligence development reaches the entire population.

Step 4 - Evaluating Governance Transparency

Trust becomes essential in an AI-driven society.

The Civilization Index examines:

- transparency of algorithmic decision systems,
- ethical oversight structures,
- citizen participation mechanisms,
- public trust indicators.

Findings:
Certain automated public services lack sufficient transparency.

Action Taken

The nation establishes:

- algorithm audit frameworks,
- explainable AI requirements,
- public accountability dashboards.

Outcome:
Citizens understand and trust intelligent governance systems.

Technology strengthens democracy rather than weakening it.

Integrated Impact of Civilization Measurement

After evaluating all dimensions, the nation gains a comprehensive understanding of its civilization health.

Instead of isolated policy decisions, leadership now sees interconnected patterns:

- environmental sustainability influences economic stability,
- education influences innovation quality,
- governance transparency influences social trust,
- workforce wellbeing influences national resilience.

The Civilization Index transforms governance into systems thinking.

Policy Evolution Through Civilization Intelligence

The findings lead to coordinated transformation:

Policy Refinement
National strategies align with Sustainable AI principles.

Educational Reform
Schools evolve from knowledge delivery to mindset development.

Responsible Innovation Incentives
Organizations receive support for ESG-aligned technology deployment.

Innovation becomes guided rather than uncontrolled.

Progress Becomes Intentional Rather Than Accidental

Traditional development models often produced unintended consequences.

The ZHA Civilization Index enables nations to design outcomes deliberately.

Leaders begin asking:

- What kind of society are we building?
- How does intelligence shape human wellbeing?
- Will our innovation remain sustainable across generations?

Civilization evolves consciously.

Long-Term National Outcomes

Over time, the nation experiences measurable transformation:

- economic resilience improves through sustainable innovation,
- environmental protection stabilizes ecosystems,
- workforce adaptability increases social stability,
- public trust strengthens governance legitimacy,
- ethical technology attracts global collaboration.

The country becomes a model of Sustainable AI civilization.

Global Ripple Effects

Other nations observe the success of ZHA alignment.

Instead of competition, cooperative learning emerges.

Countries share:

- sustainability strategies,
- governance frameworks,
- education models,
- responsible innovation practices.

Civilization progress spreads organically across borders.

The ZHA Insight

The greatest transformation of the AI era is not technological.

It is civilizational self-awareness.

When nations measure civilization health continuously, they gain the ability to guide their future consciously.

The Civilization Index enables humanity to move from:

- Reactive governance → Conscious civilization design.

17.11 The World Council as Civilization Foresight Engine

From Monitoring Intelligence to Guiding Humanity's Future

Beyond Governance, The Need for Foresight

Throughout history, governance has largely been reactive.

Societies traditionally responded after disruption occurred:

- economic crises triggered reforms,
- environmental disasters prompted regulation,
- social unrest forced political adaptation,
- technological misuse created new laws.

In the Artificial Intelligence era, this reactive model becomes insufficient.

AI systems evolve faster than human institutions.

By the time problems become visible, they may already operate at global scale.

Therefore, humanity requires something new:

not only governance but foresight.

The ZHA World Council evolves into a Civilization Foresight Engine, ensuring that humanity anticipates transformation rather than merely responding to it.

From Monitoring to Anticipation

Traditional global bodies monitor indicators:

- economic performance,
- geopolitical stability,

- environmental trends.

The ZHA World Council extends this function.

It asks deeper questions:

- How will intelligence reshape society before disruption occurs?
- What risks emerge when machines influence decision-making?
- How can civilization prepare before instability appears?

The Council becomes humanity's collective strategic intelligence system.

Core Foresight Domains

The World Council anticipates civilization-scale transitions across multiple dimensions.

1. AI Employment Transformation

The Emerging Challenge

Artificial Intelligence transforms work faster than any previous technological revolution.

AI systems increasingly perform:

- cognitive analysis,
- creative generation,
- operational decision-making,
- autonomous optimization.

Entire professions evolve simultaneously.

Without foresight:

- workforce displacement increases,
- inequality expands,
- social instability grows.

Council Foresight Role

The World Council continuously analyzes:

- global labor trends,
- emerging skill ecosystems,
- human-AI collaboration models,
- long-term employment sustainability.

Guidance Actions

The Council recommends:

- global reskilling frameworks,
- lifelong learning policies,
- transition income models,
- human-centered job redesign.

Outcome:
Automation becomes human augmentation, not human replacement.

Civilization adapts before disruption occurs.

2. Climate Migration Risks

The Emerging Challenge

Climate change increasingly drives population movement due to:

- rising sea levels,
- water scarcity,

- extreme weather,
- ecosystem collapse.

AI accelerates environmental prediction, but governance often reacts too late.

Council Foresight Role

Using Sustainable AI intelligence models, the Council anticipates:

- migration corridors,
- urban pressure zones,
- resource stress regions,
- humanitarian risk scenarios.

Guidance Actions

The Council facilitates:

- cooperative international planning,
- sustainable city expansion strategies,
- climate-resilient infrastructure investment,
- shared humanitarian response frameworks.

Outcome:
Migration becomes managed adaptation rather than crisis response.

Civilization preserves dignity during environmental transition.

3. Digital Inequality

The Emerging Challenge

AI creates unprecedented opportunity but also new inequality.

Nations and communities lacking access to:

- AI education,
- computing infrastructure,
- digital literacy,
- innovation ecosystems
- risk exclusion from the intelligent economy.

Without foresight:

- intelligence wealth concentrates,
- technological colonialism emerges,
- social fragmentation increases.

Council Foresight Role

The World Council monitors:

- AI access distribution,
- education equity,
- infrastructure readiness,
- digital participation levels.

Guidance Actions

Recommendations include:

- global AI literacy initiatives,
- open innovation collaboration,
- shared research access,
- equitable technology deployment frameworks.

Outcome:

AI becomes a global public good, not a privilege of few institutions.

Civilization evolves inclusively.

4. Autonomous Governance Implications

The Emerging Challenge

AI increasingly participates in governance itself:

- automated policy analysis,
- predictive policing,
- algorithmic welfare allocation,
- autonomous infrastructure management.

For the first time, governance decisions may be partially shaped by intelligent systems.

Without foresight:

- accountability becomes unclear,
- democratic trust weakens,
- algorithmic authority expands unchecked.

Council Foresight Role

The World Council evaluates:

- ethical limits of automation,
- human oversight requirements,
- explainability standards,
- societal acceptance thresholds.

Guidance Actions

The Council promotes:

- Human-in-the-Loop governance,
- transparent decision systems,
- ethical AI certification standards,
- citizen participation mechanisms.

Outcome:
AI strengthens governance legitimacy rather than replacing human agency.

Civilization Foresight Methodology

The ZHA World Council operates through a continuous foresight cycle:

Observe
collect global data signals.

Anticipate
model long-term scenarios.

Evaluate
assess civilizational risks and opportunities.

Guide
publish recommendations and standards.

Align
support institutions in coordinated adaptation.

Civilization becomes capable of learning ahead of time.

Humanity's Early Warning System

In practical terms, the World Council functions like:

- a planetary risk radar,
- a sustainability compass,

- a long-term intelligence advisor.

It identifies weak signals before they become crises.

Examples include:

- emerging AI misinformation risks,
- psychological effects of human-AI interaction,
- unintended environmental impacts of compute expansion,
- societal dependency on automated systems.

The Council does not control nations or organizations.

It illuminates direction.

From Reactive Civilization to Proactive Civilization

Historically:
Civilizations reacted to change.

In the AI era:
Civilizations must anticipate change.

The ZHA World Council enables humanity to transition from:

- Reactive Survival → Conscious Evolution.

The Deeper ZHA Insight

The greatest danger of Artificial Intelligence is not intelligence itself.

It is intelligence evolving faster than wisdom.

Foresight bridges this gap.

When humanity anticipates technological consequences collectively:

- crises reduce,
- trust strengthens,
- innovation stabilizes,
- civilization matures.

17.12 Sustainable AI as Coordinated Global Intelligence

From Fragmented Innovation to Collective Human Evolution

The Need for Coordinated Intelligence

Human civilization has reached a historic threshold.

For the first time in history, intelligence itself is becoming a shared global infrastructure.

Artificial Intelligence systems now:

- learn continuously,
- influence billions simultaneously,
- reshape economies instantly,
- guide decisions across governments, industries, and communities.

Yet humanity still operates largely through fragmented intelligence:

- organizations innovate independently,
- nations regulate separately,
- institutions pursue isolated priorities,
- societies experience uneven technological outcomes.

This creates a fundamental imbalance:

Global Intelligence without Global Coordination.

Sustainable AI introduces a new paradigm:

Intelligence must evolve as a coordinated global system serving humanity collectively.

The ZHA World Council and the Civilization Index together enable this transformation.

Sustainable AI as Coordinated Global Intelligence

Under the ZHA framework, Sustainable AI is not merely ethical technology.

It becomes Coordinated Global Intelligence — a living feedback system connecting:

- innovation,
- governance,
- societal outcomes,
- continuous learning.

Human civilization begins to operate like an intelligent organism capable of self-reflection and adaptation.

The ZHA Civilization Feedback Loop

The interaction between the World Council and the Civilization Index forms a continuous evolutionary cycle.

1. Principles Guide Innovation

Every intelligent system begins with values.

The ZHA World Council establishes foundational civilizational principles:

- Human-centric purpose,
- sustainability mindset,

- ESG responsibility,
- long-term societal wellbeing,
- planetary stewardship.

These principles act as navigation coordinates for innovation.

Example: AI Product Development

A global company developing healthcare AI aligns with ZHA principles:

Instead of optimizing only accuracy:

- patient dignity becomes priority,
- energy efficiency becomes requirement,
- accessibility becomes design goal.

Innovation begins ethically aligned.

Technology develops in the right direction from the start.

2. Innovation Produces Societal Outcomes

Once deployed, AI systems influence real civilization conditions:

- employment structures change,
- education models evolve,
- environmental impacts shift,
- governance processes transform.

Innovation is no longer neutral.

Every AI system generates civilizational consequences.

Example: Smart Agriculture AI

AI introduced into farming leads to:

- improved crop yield,
- reduced water consumption,
- better farmer income stability.

But it may also:

- alter labor patterns,
- require reskilling,
- change rural economic structures.

Innovation produces multidimensional outcomes.

3. Outcomes Are Measured

Here the ZHA Civilization Index becomes essential.

The Index evaluates whether innovation strengthens civilization across five dimensions:

- Environmental sustainability,
- Social inclusion,
- Governance transparency,
- Ethical innovation,
- Human wellbeing.

Measurement transforms intuition into insight.

Example: Measuring AI Deployment

A nation evaluates its AI ecosystem:

- Are AI data centers increasing carbon emissions?
- Are workers supported during automation?

- Do citizens trust algorithmic decisions?

Civilization health becomes measurable.

4. Measurement Informs Future Governance

Measurement alone is insufficient.

Insight must guide action.

The ZHA World Council analyzes Civilization Index findings to:

- refine global recommendations,
- update ethical standards,
- anticipate emerging risks,
- encourage collaborative solutions.

Governance evolves dynamically rather than through static regulation.

Example: AI Energy Consumption Alert

Civilization Index data shows rising global compute energy usage.

The World Council responds by recommending:

- energy-efficient model standards,
- shared green computing infrastructure,
- sustainability certification for AI platforms.

Governance adapts intelligently.

Civilization Learns Continuously

This cycle creates something unprecedented:

A learning civilization.

Instead of repeating crises, humanity begins to learn from its own technological behavior.

The system operates like a biological feedback loop:

Civilization becomes capable of self-correction.

AI Becomes a Partner in Human Evolution

Historically, technology served as a tool.

In the Sustainable AI era, intelligence becomes a collaborative partner.

AI assists humanity to:

- understand complex systems,
- model long-term futures,
- predict risks early,
- coordinate global responses.

Example: Climate Intelligence Partnership

AI models analyze planetary data to forecast ecological risk.

Human leaders interpret results using ethical frameworks.

Together:

- AI provides intelligence,
- humanity provides wisdom.

Evolution becomes cooperative.

From Fragmented Intelligence to Collective Intelligence

Without coordination:

- innovation competes,
- knowledge fragments,
- risks multiply.

With ZHA alignment:

- intelligence synchronizes,
- institutions collaborate,
- progress stabilizes.

Human civilization transitions from:

- Independent Actors → Coordinated Global Intelligence Network

The Emergence of Civilization Intelligence

The ZHA framework introduces a new concept:

Civilization Intelligence

A state where humanity collectively:

- observes itself,
- learns from outcomes,
- adapts responsibly,
- evolves consciously.

Artificial Intelligence amplifies this capability.

Humanity moves from accidental progress to intentional evolution.

Real-World Scenario - Coordinated Global Intelligence

Imagine a future aligned with ZHA principles:

Universities teach sustainability mindset globally.

Companies build AI guided by ESG frameworks.

Governments align policies using Civilization Index insights.

Communities participate in governance discussions.

Feedback flows continuously across society.

Innovation becomes harmonized rather than chaotic.

Civilization operates as a coordinated intelligence ecosystem.

The ZHA Insight

The future challenge is no longer creating intelligence.

It is coordinating intelligence wisely.

When principles guide innovation,
when outcomes are measured,
when governance learns continuously, humanity becomes capable of guiding its own evolution.

17.13 Finally - Measuring What Humanity Truly Values

Civilizations become what they measure.

If humanity measures only profit,
it produces exploitation.

If humanity measures power,
it produces competition.

If humanity measures wellbeing,
sustainability,
and responsibility,
it produces conscious civilization.

The Civilization Index ensures humanity measures what truly matters.

CHAPTER XVIII

Chapter 18: Founder Leadership Philosophy

Introduces institutional structures designed to scale Sustainable AI globally through governance, research, and collaborative leadership.

Introduction - Leadership in the Age of Intelligent Civilization

Every civilization movement begins not with technology, but with a shift in leadership consciousness.

The ZHA Sustainability Mindset did not emerge as a theoretical framework alone.
It emerged from a deeper realization:

The greatest crisis of the AI age is not technological capability — it is leadership maturity.

Artificial Intelligence amplifies human intention.

If leadership is driven by:

- greed,
- short-term thinking,
- competition without responsibility,
- AI accelerates instability.

If leadership is guided by:

- consciousness,
- sustainability,
- compassion,
- long-term stewardship,

AI becomes humanity's greatest evolutionary partner.

This chapter introduces the leadership philosophy behind the ZHA Methodology, developed by its founder, Karki Ashokkumar, and explains how leadership itself becomes civilization infrastructure.

18.1 The Origin of the ZHA Sustainability Mindset

The ZHA methodology originated from observing a recurring pattern across industries and societies:

- Technological progress increased,
- but human fulfillment decreased.

Organizations became more efficient.

Societies became more connected.

Yet individuals experienced:

- rising anxiety,
- ecological imbalance,
- social fragmentation,
- loss of meaning in work and life.

The insight was profound:

Humanity was advancing technologically without evolving consciously.

The ZHA Sustainability Mindset emerged as a response - a framework integrating:

- inner transformation,
- social responsibility,
- environmental stewardship,

- technological wisdom.

Leadership therefore begins not from authority but from self-awareness.

18.2 Leadership Philosophy - From Control to Conscious Stewardship

Traditional leadership models emphasize:

- hierarchy,
- control,
- performance metrics,
- competitive advantage.

The ZHA leadership philosophy introduces a fundamental shift:

Leadership evolves from managing systems to guiding civilization outcomes.

Core Principle

Leadership is not the ability to command people.
Leadership is the ability to align intelligence with humanity's future.

18.3 The Founder's Insight - Sustainability Begins Within

The ZHA methodology is built upon a foundational belief:

Civilization problems originate from mindset imbalance.

Environmental crises,
economic inequality,
technological misuse

all emerge from human behavior patterns.

Therefore:

Sustainable AI requires Sustainable Humans.

The founder's philosophy emphasizes:

- inner sufficiency before external growth,
- awareness before automation,
- wisdom before innovation.

Leadership transformation precedes technological transformation.

18.4 Leadership as a Civilization Responsibility

In the AI era, leaders influence outcomes at planetary scale.

A single decision can impact:

- millions of workers,
- global ecosystems,
- democratic stability,
- collective human behavior.

Thus leadership evolves into civilization stewardship.

Example - AI Deployment Decision

A leader approving large-scale automation faces two paths:

Traditional Decision

- maximize efficiency,
- reduce workforce,
- increase profit margins.

ZHA Leadership Decision

- integrate reskilling programs,
- redesign human-AI collaboration,
- protect dignity while improving productivity.

Leadership becomes ethical navigation.

18.5 The ZHA Leadership Model

The ZHA methodology defines leadership across five evolutionary stages.

Stage 1 - Personal Awareness

Leader asks:

- Who am I serving?
- What impact do my decisions create?

Focus:
Self-awareness and responsibility.

Stage 2 - Organizational Alignment

Leader aligns:

- culture,
- values,
- SDLC practices,
- innovation strategy.

Outcome:
Responsible organizations.

Stage 3 - Societal Contribution

Leader considers:

community wellbeing,
social inclusion,
environmental responsibility.
Outcome:
Organizations become social contributors.

Stage 4 - Sustainable Innovation Leadership

Leader guides:

- ESG-driven innovation,
- ethical AI adoption,
- human-centered product development.

Outcome:
Sustainable progress.

Stage 5 - Civilization Stewardship

Leader operates beyond organization boundaries.

Focus:

- future generations,
- planetary balance,
- collective intelligence.

Outcome:
Civilization leadership.

18.6 Leadership and Sustainable AI Transformation

The ZHA founder recognized that AI transformation fails when treated purely as a technology initiative.

Most organizations begin AI adoption by asking:

- Which tools should we implement?
- Which algorithms should we deploy?

ZHA leadership begins differently:

- Why are we building intelligence?

The sequence changes:

- Align Purpose
- Transform Mindset
- Design Responsible Systems
- Deploy Sustainable AI

Leadership becomes the starting point of transformation.

Example - Enterprise AI Transformation

Without ZHA leadership:

- AI implemented for cost reduction,
- workforce anxiety increases,
- trust declines.

With ZHA leadership:

- employees trained as AI collaborators,
- ethical frameworks established,
- innovation culture strengthens.

Technology succeeds because leadership matured first.

18.7 Collaborative Leadership - Beyond Individual Heroism

The founder philosophy rejects the myth of the heroic individual leader.

Civilization challenges are too complex for individual brilliance alone.

ZHA promotes collective leadership:

- students,
- professionals,
- institutions,
- communities.

Leadership becomes a distributed capability across society.

Example - ZSPC Ecosystem

Students practice sustainability mindset.
Professionals mentor innovation projects.
Organizations support ESG initiatives.
Leadership emerges collectively.

18.8 Leadership and the Sustainability Mindset Principles

The twelve Sustainability Mindset Principles act as leadership operating guidelines.

Leaders practice:

- Feeling sufficient → reduces exploitation.
- Collaboration → strengthens inclusion.
- Long-term thinking → protects future generations.
- Continuous learning → enables adaptation.
- Encouragement → builds human potential.

Leadership becomes lived behavior rather than organizational policy.

18.9 The Founder's Philosophy - Sustainable AI with Human-Centric

Purpose

At the heart of ZHA leadership lies a defining realization:

- Responsible AI is not enough.
- Responsible AI prevents harm.

Sustainable AI with Human-Centric Purpose creates flourishing.

The difference is transformational.

Leadership shifts from fear-based regulation to purpose-driven creation.

18.10 Building Leaders for the Intelligent Civilization

The ZHA methodology aims to cultivate a new leadership archetype:

The Sustainable AI Leader

A leader who:

- understands technology deeply,
- respects humanity profoundly,
- thinks systemically,
- acts ethically,
- serves future generations.

Such leaders become:

- educators,
- innovators,
- policymakers,
- civilization designers.

Example - Education Leadership Transformation

A school adopting ZHA principles:

- teaches AI literacy,
- embeds sustainability mindset,
- encourages community problem solving.

Students graduate not just as professionals but as responsible innovators.

Leadership pipeline becomes civilizational.

18.11 Institutionalizing Leadership Philosophy

The founder's philosophy scales through institutions:

- ZHA Global Institute — leadership education.
- ZHA World Council — civilization stewardship.
- ZSPC Certification Ecosystem — practitioner development.
- Industry Consortium — responsible innovation adoption.

Leadership becomes systemic rather than personality-driven.

18.12 The ZHA Insight - Leadership Determines Civilization Destiny

History reveals a consistent truth:

Civilizations rise or fall based on leadership maturity.

Technology amplifies leadership decisions.

Artificial Intelligence amplifies them exponentially.

Therefore:

The future of humanity will not be decided by AI capability, but by the consciousness of those who guide it.

The ZHA methodology represents leadership evolution itself:

- From power → responsibility
- From management → stewardship
- From intelligence → wisdom

Chapter Reflection

The ZHA Sustainability Mindset is more than a framework.

It is a leadership philosophy for the Intelligent Age.

Karki Ashokkumar's foundational insight reframes AI transformation:

Humanity does not merely need smarter machines.

Humanity needs wiser leaders.

Through Sustainable AI with Human-Centric Purpose, leadership becomes the force that aligns innovation, governance, education, and society toward a conscious civilization.

The future belongs not to those who control intelligence, but to those who guide intelligence with purpose, compassion, and sustainability.

PART V - Global Governance and Thought Leadership

CHAPTER XIX

Chapter 19: ZHA Global Think Tank

Shaping the Future of Sustainable Innovation and Responsible AI Governance

Introduction - The Age of Intelligence Requires Collective Wisdom

Human civilization has entered an unprecedented phase.

For the first time in history:

- intelligence is programmable,
- decision-making is automated,
- knowledge evolves continuously,
- innovation scales globally within seconds.

Artificial Intelligence no longer belongs to laboratories or corporations alone.
It influences governance, education, economies, culture, and human behavior itself.

Yet one fundamental challenge emerges:?
Humanity possesses powerful intelligence systems but lacks unified wisdom to guide them.

Technology advances rapidly.
Governance reacts slowly.
Research remains fragmented.
Leadership operates in silos.

The ZHA Global Think Tank emerges as humanity's response, a platform designed to transform intelligence into collective civilization wisdom.

19.1 From Knowledge Production to Civilization Intelligence

Traditional research institutions focus on producing knowledge.

ZHA reframes research itself.

Knowledge alone is insufficient.

Humanity requires actionable wisdom.

Example: Climate AI Research

Traditional model:

publish climate prediction studies.

ZHA Think Tank model:

- integrate research,
- advise governments,
- guide industry implementation,
- educate communities.

Research becomes civilization stewardship.

19.2 Thought Leadership in the Age of Sustainable AI

Thought leadership in the AI era must evolve beyond expert opinion.

ZHA defines thought leadership as:
The ability to anticipate civilization consequences before technology reshapes society.

True thought leadership integrates:

- technology,
- ethics,
- sustainability,

- governance,
- human psychology.

Example: Generative AI

A traditional response:

- optimize models,
- compete for performance.

ZHA thought leadership asks:

- How will generative AI reshape human creativity?
- What happens to education?
- How does society maintain authenticity?

Thought leadership becomes foresight.

19.3 Core Functions of the ZHA Global Think Tank

The Think Tank operates across five strategic domains.

1. Sustainable AI Policy Intelligence

Develops guidance for:

- ethical AI deployment,
- ESG-aligned innovation,
- sustainable governance models.

Supports governments without replacing sovereignty.

2. Civilization Foresight Research

Anticipates long-term impacts:

- AI employment transformation,
- digital inequality,
- autonomous governance,
- human-AI coexistence.

Civilization becomes proactive rather than reactive.

3. Responsible Innovation Advisory

Helps organizations design:

- human-centered products,
- sustainable SDLC practices,
- ethical innovation ecosystems.

Innovation aligns with civilization wellbeing.

4. Global Collaboration Platform

Connects:

- universities,
- industries,
- policymakers,
- civil society.

Competition evolves into cooperative intelligence.

5. Leadership Development

Cultivates Sustainable AI leaders capable of guiding intelligent civilization.

Leadership becomes foresight-driven.

19.4 The Think Tank as a Bridge Between Worlds

Modern civilization suffers from fragmentation.

The ZHA Global Think Tank integrates these worlds.

It becomes a bridge between knowledge and responsibility.

Example - Smart City Development

Without integration:

- technology vendors deploy systems independently,
- citizens lack participation,
- governance reacts after deployment.

With ZHA Think Tank guidance:

- citizen engagement included,
- ESG metrics defined,
- sustainability evaluated before rollout.

Technology becomes socially aligned.

19.5 Global Collaboration as Civilization Strategy

The AI era makes isolation impossible.

A model trained in one country affects users worldwide instantly.

Therefore collaboration becomes survival.

The ZHA Think Tank promotes:

- shared research ecosystems,

- open sustainability standards,
- collaborative governance dialogues,
- cross-cultural innovation exchange.

Example - Pandemic Intelligence Systems

Global health AI systems improve dramatically when nations share data responsibly rather than compete.

Collective intelligence protects civilization.

19.6 Sustainable Innovation Research Domains

The Think Tank prioritizes interdisciplinary research areas.

Sustainable AI Architectures
Energy-efficient intelligence systems.

Human AI Collaboration Models
AI enhances human creativity rather than replaces it.

Climate Intelligence
AI supports planetary resilience.

Ethical Algorithm Design
Bias prevention and fairness systems.

Wellbeing Technology
AI improves quality of life.

Research becomes civilization design.

19.7 The ZHA Think Tank Governance Model

From Centralized Authority to Collaborative Intelligence

Introduction : Governance in the Age of Intelligent Systems

Artificial Intelligence has fundamentally changed how decisions influence society.

In earlier eras:

- decisions affected organizations,
- policies affected nations,
- innovations affected industries.

Today, AI decisions affect entire civilizations simultaneously.

A recommendation algorithm influences billions of people.
A climate model impacts global policy.
An autonomous system shapes economic futures.

Traditional governance models, built around centralized authority, struggle to manage such distributed intelligence.

The ZHA Global Think Tank introduces a new governance philosophy:
Intelligence must be guided by collective wisdom, not isolated power.

Governance therefore evolves from command structures to collaborative intelligence ecosystems.

19.7.1 The Limits of Traditional Governance Models

Historically, governance followed hierarchical patterns:

- Centralized Model
- authority concentrated at the top,
- experts advise leaders,
- decisions flow downward.

This model worked when systems were predictable.

However, AI ecosystems are:

- dynamic,
- self-learning,
- globally interconnected,
- socially sensitive.

No single institution possesses sufficient perspective to govern intelligent systems alone.

Example: Social Media Algorithms

Centralized decision:
optimize engagement metrics.

Unintended consequence:

- misinformation amplification,
- social polarization,
- mental health impacts.

Failure occurred not because of bad technology, but because decision perspectives were limited.

19.7.2 Distributed Participation - The ZHA Governance Principle

The ZHA Think Tank replaces centralized authority with distributed participation.

Governance becomes a living ecosystem where multiple knowledge domains collaborate continuously.

Decision-making integrates perspectives from:

- science,
- society,

- environment,
- economics,
- ethics,
- human wellbeing.

Authority shifts from power ownership to responsibility sharing.

Core Insight

No single discipline can govern intelligence.
Civilization governance requires interdisciplinary wisdom.

19.7.3 Structure of Distributed Councils

The ZHA Think Tank operates through interconnected councils representing civilization stakeholders.

Each council contributes a unique intelligence layer.

1. Sustainability Researchers

Role
Ensure innovation aligns with planetary limits.

They evaluate:

- environmental impact of AI infrastructure,
- energy consumption,
- resource sustainability,
- biodiversity implications.

Example:

Before deploying large AI data centers:

- researchers assess carbon footprint,

- recommend renewable energy integration,
- guide sustainable compute architecture.

Technology development becomes ecological stewardship.

2. AI Scientists and Technologists

Role
Provide technical insight into emerging intelligence systems.

They analyze:

- algorithm capability,
- system risks,
- bias mitigation,
- safe deployment practices.

Example:

When autonomous decision systems are introduced in healthcare:

- AI scientists validate model reliability,
- ensure explainability,
- define safety thresholds.
- Innovation proceeds responsibly.

3. Educators and Academic Leaders

Role
Shape long-term human capability.

They guide:

- AI literacy programs,
- sustainability mindset education,

- ethical innovation training.

Example:

Think Tank recommendations influence universities to integrate:

- AI engineering,
- ethics,
- environmental science,
- into unified curricula.

Future innovators develop responsible thinking before entering industry.

4. Industry Innovators

Role
Translate ideas into real-world implementation.

They contribute:

- operational feasibility,
- business transformation insights,
- product lifecycle realities.

Example:

A manufacturing company adopting AI automation collaborates with the Think Tank to:

- reskill workers,
- reduce emissions,
- maintain productivity.

Industry becomes a civilization partner rather than a disruption driver.

5. Policymakers and Governance Experts

Role
Align public policy with technological evolution.

They help develop:

- adaptive regulations,
- ESG governance standards,
- international cooperation frameworks.

Example:

Instead of banning emerging AI technologies reactively, policymakers co-design ethical deployment frameworks with scientists and communities.

Governance becomes anticipatory.

6. Social Leaders and Civil Society Representatives

Role
Represent human experience and societal impact.

They ensure governance considers:

- cultural diversity,
- public trust,
- social equity,
- community wellbeing.

Example:

During deployment of AI surveillance systems:

- community leaders assess privacy concerns,

- inclusion risks are addressed,
- citizen trust becomes a design requirement.

Human dignity remains central.

19.7.4 Collaborative Intelligence Decision Process
The ZHA Think Tank follows a collaborative decision cycle.

Step 1 - Multidisciplinary Insight Gathering
Each council analyzes innovation from its perspective.

Step 2 - Collective Dialogue
Stakeholders exchange insights openly.

Step 3 - Impact Synthesis
Environmental, social, technological, and governance impacts integrated.

Step 4 - Responsible Recommendation
Guidelines produced balancing innovation and sustainability.

Step 5 - Continuous Learning
Outcomes monitored and governance evolves.

Example: AI Education Platform Deployment

Traditional model:
tech company launches platform independently.

ZHA Think Tank model:

- educators assess learning impact,
- scientists validate fairness,
- policymakers ensure data protection,
- social leaders evaluate accessibility.

Result:

Responsible innovation with societal trust.

19.7.5 Authority Becomes Shared Wisdom

In the ZHA governance model:

Authority no longer means control.

Authority means earned trust through collective intelligence.

Leadership transforms into:

- facilitation,
- stewardship,
- coordination.

No council dominates.
All perspectives contribute.

Governance Transformation

19.7.6 Benefits of Collaborative Governance

1. Better Decision Quality
Multiple perspectives reduce blind spots.

2. Increased Public Trust
Citizens see representation in governance processes.

3. Faster Ethical Adaptation
Governance evolves alongside innovation.

4. Sustainable Innovation
Technology aligns naturally with ESG principles.

Real Example: Climate Intelligence Systems

Global AI climate platforms governed collaboratively:

- scientists provide models,
- policymakers enact strategies,
- industry implements solutions,
- communities adapt locally.

Civilization acts as a coordinated organism.

19.7.7 Governance as Living Intelligence

The ZHA Think Tank treats governance not as static rules but as a learning system.

Governance continuously evolves because:

- AI evolves,
- society evolves,
- environmental conditions evolve.

The Think Tank becomes a living intelligence network guiding humanity's relationship with technology.

19.7.8 The ZHA Insight - Wisdom Must Scale with Intelligence

Humanity has successfully scaled:

- communication,
- computation,
- automation.

Now humanity must scale wisdom.

Artificial Intelligence multiplies human capability.

The ZHA Think Tank ensures wisdom multiplies alongside it.

19.8 Thought Leadership as Global Responsibility

In the AI era, influence carries ethical responsibility.

Thought leaders shape:

- technological direction,
- societal expectations,
- innovation culture.

ZHA promotes responsible thought leadership:

Leaders must ask:

- Who benefits?
- Who is excluded?
- What future are we creating?

Thought leadership becomes moral leadership.

19.9 The Think Tank as Civilization Foresight Engine

Beyond research and advice, the ZHA Global Think Tank performs a critical function:

anticipating civilization risks before they emerge.

Areas of foresight include:

- AI governance evolution,
- automation and human purpose,
- climate migration,
- digital identity ethics,
- AI and democracy.

Humanity gains an early warning system.

19.10 Sustainable AI as Collective Intelligence

The Think Tank embodies a new principle:

AI transformation cannot be guided by isolated genius.

It requires collective intelligence.

The feedback loop becomes:
Sustainability principles guide innovation.
Innovation produces societal outcomes.

Outcomes are evaluated through ESG metrics.

Insights guide future governance.

Civilization learns continuously.

19.11 The ZHA Insight - Thought Leadership Shapes Civilization Direction

Every era has been defined by ideas before institutions followed.

The Renaissance was guided by thinkers.

The Enlightenment by philosophers.

The Digital Age by technologists.

The AI era must be guided by sustainability thinkers.

The ZHA Global Think Tank represents the evolution of thought leadership itself:

- From expertise → wisdom

- From innovation → responsibility
- From intelligence → civilization stewardship

Chapter Reflection

Artificial Intelligence will define humanity's future. But technology alone cannot decide that future.

Only collective wisdom can.

The ZHA Global Think Tank transforms thought leadership into civilization leadership ensuring Sustainable AI with Human-Centric Purpose guides innovation, governance, and global collaboration.

Through this institution, humanity moves from fragmented progress toward a consciously guided intelligent civilization.

CHAPTER XX

Chapter 20: World Summit and Founders Platform

"Global Alignment for Sustainable AI with Human-Centric Purpose"

Introduction : From Ideas to Global Civilization Action

Throughout this book, the ZHA journey has evolved across multiple layers:

- Sustainability Mindset
- AI Transformation
- Responsible Innovation
- Civilization Governance
- Institutional Architecture
- Global Thought Leadership

Each chapter builds toward one inevitable realization:

Civilization transformation cannot remain theoretical.

It must become a shared global practice.

Ideas alone do not change civilization.

Movements do.
Institutions do.
Collaborative leadership does.

The ZHA World Summit and Founders Platform represent the moment where thought leadership transforms into collective civilization action.

It is the global convergence point for humanity's transition toward:

Sustainable AI with Human-Centric Purpose.

20.1 Why Humanity Needs a Global World Summit in the AI Era

Artificial Intelligence has created a paradox.

Never before has humanity been so connected technologically, yet so fragmented institutionally.

Today:

- researchers innovate independently,
- companies deploy AI competitively,
- governments regulate locally,
- educators evolve unevenly.

AI systems operate globally, but decision-making remains fragmented.

This creates risks:

- ethical inconsistency,
- technological inequality,
- ESG instability,
- declining public trust.

History shows that civilization advances when humanity gathers to align vision.

Examples include:

- international scientific congresses,
- climate summits,
- global economic forums.

The AI era now demands a summit dedicated not to politics or markets, but to civilization stewardship.

The ZHA World Summit fulfills this role.

20.2 The Purpose of the ZHA World Summit

The ZHA World Summit serves as humanity's collaborative platform for guiding intelligent civilization.

Core Mission
To align global leaders around a shared commitment:

- Sustainable innovation,
- Responsible AI governance,
- Human-centered development,
- Planetary stewardship.

The summit does not impose authority.

It creates alignment.

Key Objectives

- Establish global dialogue on Sustainable AI.
- Connect institutions across sectors.
- Share responsible innovation practices.
- Develop civilization-scale foresight.
- Inspire collaborative global action.

The Summit becomes a living forum for civilization intelligence.

20.3 From Conference to Civilization Platform

Traditional conferences focus on:

- presentations,
- networking,
- announcements.

The ZHA World Summit redefines the concept.

It is not an event.

It is an ongoing global alignment mechanism.

The Summit becomes a recurring moment where humanity reflects on the direction of intelligence itself.

20.4 The Founders Platform - Leadership for the Intelligent Civilization

At the heart of the Summit lies the ZHA Founders Platform.

The Founders Platform recognizes a new leadership archetype emerging globally:

- innovators,
- educators,
- entrepreneurs,
- policymakers,
- social leaders,
- sustainability practitioners.

These individuals are not merely organizational founders.

They are civilization founders, people shaping how humanity lives with intelligent systems.

Founders Platform Purpose

To unite global change-makers who commit to:

- building Sustainable AI ecosystems,
- guiding ethical innovation,
- nurturing sustainability mindset leadership,
- protecting civilization continuity.

Leadership becomes shared responsibility.

20.5 The Rise of the Civilization Founder

Historically, founders built companies.

In the AI era, founders shape systems that influence humanity itself.

Examples include:

- creators of digital platforms,
- builders of AI infrastructure,
- architects of educational ecosystems,
- designers of governance frameworks.

The ZHA Founders Platform reframes the role:

A Founder is a steward of civilization impact.

Example: Responsible AI Startup Founder

Traditional founder focus:

- market valuation,
- rapid scale,
- competitive advantage.

ZHA Founder perspective:

- ethical data use,

- inclusive access,
- environmental sustainability,
- societal wellbeing.

Success expands beyond profit toward civilization value creation.

20.6 Structure of the ZHA World Summit

The Summit operates through interconnected forums.

1. Sustainable AI Leadership Forum
Global executives and innovators align AI transformation strategies with ESG principles.

Focus:

- Responsible SDLC practices,
- Sustainable product ecosystems,
- Human-AI collaboration models.

2. Research & Thought Leadership Forum
Academia and think tanks present civilization-focused research.

Topics:

- climate intelligence,
- AI ethics,
- social impact modeling,
- future governance systems.

3. Education & Youth Civilization Forum
Students and educators participate actively.

Purpose:

- empower next-generation sustainability practitioners,
- integrate ZSPC learning ecosystems globally.

Civilization continuity begins with youth.

4. Policy and Governance Forum
Policymakers collaborate on anticipatory governance frameworks.

Outcome:

- shared ethical standards,
- international cooperation models,
- sustainable AI policies.

5. Industry Collaboration Forum
Corporations exchange responsible innovation practices.

Competition evolves into cooperative progress.

20.7 The World Summit as a Global Alignment Engine

The ZHA World Summit functions as an alignment engine connecting:

- education,
- industry,
- governance,
- society,
- research.

Instead of isolated transformation efforts, humanity moves toward coordinated evolution.

Example: Global AI Education Alignment

Through Summit collaboration:

- universities adopt sustainability mindset curricula,
- companies define ethical hiring practices,
- governments support AI literacy policies.

Transformation scales globally.

20.8 Thought Leadership Becomes Collective Leadership

The Summit democratizes thought leadership.

Wisdom emerges not from a single authority but from collective dialogue.

Participants co-create:

- foresight reports,
- sustainability frameworks,
- innovation guidelines,
- civilization commitments.

Thought leadership evolves into shared intelligence.

20.9 Sustainable Innovation Through Global Collaboration

Innovation challenges today are planetary:

- climate change,
- workforce transformation,
- digital inequality,
- ethical AI deployment.

No organization or nation can solve them alone.

The ZHA World Summit enables collaborative innovation ecosystems where:

- companies share sustainability practices,
- researchers collaborate across borders,
- communities contribute lived experience.

Innovation becomes civilization cooperation.

20.10 The Founders Platform as a Global Leadership Network

The Founders Platform extends beyond the Summit.

It forms a continuous global network.

Members collaborate year-round to:

- mentor emerging leaders,
- launch sustainable initiatives,
- guide AI governance evolution,
- support civilization-scale innovation.

Leadership becomes an interconnected ecosystem rather than isolated influence.

20.11 From Leadership Recognition to Responsibility Commitment

The ZHA Founders Platform does not merely celebrate success.

It recognizes responsibility.

Founders commit to:

- ethical innovation,
- sustainability integration,
- long-term societal contribution.

Recognition becomes a pledge to humanity's future.

20.12 World Summit Outcomes - Civilization Impact

The Summit generates tangible outcomes:

- global Sustainable AI declarations,
- cross-sector research collaborations,
- ESG-aligned innovation standards,
- leadership fellowship programs,
- civilization foresight agendas.

The event becomes a catalyst for continuous global progress.

Example: Sustainable AI Deployment Initiative

Following Summit discussions:

- organizations adopt shared energy-efficient AI standards,
- research institutions develop open sustainability benchmarks,
- governments align incentives.

Collective decisions produce measurable civilization benefit.

20.13 The ZHA Insight - Civilization Evolves Through Gathering

Human civilization has always advanced through collective gatherings:

- ancient assemblies shaped law,
- scientific congresses accelerated discovery,
- global summits aligned cooperation.

The ZHA World Summit continues this tradition for the Intelligent Age.

It becomes humanity's forum for answering the defining question:

How should intelligence serve civilization?

20.14 Toward a Human-Centric Intelligent Civilization
The World Summit and Founders Platform represent the culmination of the ZHA vision.

They transform:

- philosophy into practice,
- leadership into stewardship,
- innovation into sustainability,
- intelligence into civilization evolution.

Sustainable AI with Human-Centric Purpose becomes not merely an idea — but a global movement.

Chapter Reflection

Artificial Intelligence will shape the future of humanity. The ZHA World Summit ensures humanity shapes Artificial Intelligence in return.

Through global collaboration, shared leadership, and responsible innovation, the Summit establishes a new model of civilization governance one where intelligence advances with wisdom, compassion, and sustainability.

The future of civilization will not be decided by isolated innovation.

It will be decided by humanity's ability to gather, align, and act together.

The ZHA World Summit marks that beginning.

CHAPTER XXI

Chapter 21: Global Thought Leadership Strategy

"Guiding Humanity Toward Sustainable AI with Human-Centric Purpose"

Introduction - The Battle for the Future Is a Battle of Ideas

Every civilization has been shaped not first by technology or economics — but by ideas.

The Renaissance was driven by intellectual awakening.

The Enlightenment was guided by philosophical thought leadership.

The Digital Age was led by technological visionaries.

The Artificial Intelligence era introduces a new reality:

The future of civilization will be determined by those who shape how humanity understands intelligence itself.

Artificial Intelligence amplifies knowledge creation, decision-making, and societal influence.
But intelligence without guiding philosophy risks accelerating instability.

Humanity therefore requires global thought leadership capable of aligning innovation with sustainability, ethics, and human purpose.

The ZHA Global Thought Leadership Strategy emerges as a framework to guide this transformation.

21.1 The Purpose of the ZHA Global Thought Leadership Strategy
The strategy aims to position Sustainable AI with Human-Centric Purpose as a global reference framework.

Strategic Objectives

Establish Sustainable AI as the global benchmark for AI transformation.

Guide organizations toward responsible innovation.

Influence policy discussions internationally.

Educate future sustainability leaders.

Align global institutions toward civilization stewardship.

The strategy transforms ZHA from a framework into a global intellectual movement.

21.2 Thought Leadership Pillars of ZHA

The strategy operates across five integrated pillars.

Pillar 1 — Knowledge Creation

ZHA generates original intellectual frameworks:

- Sustainability Mindset methodology,
- AI ESG governance models,
- Sustainable SDLC frameworks,
- Civilization governance principles.

Example: Publishing research linking AI energy efficiency with ESG outcomes influences corporate transformation globally.

Knowledge becomes civilization guidance.

Pillar 2 - Education Leadership

Thought leadership scales through education.

ZHA promotes:

- sustainability mindset curricula,
- AI ethics education,
- interdisciplinary learning models.

Example: Universities adopting ZHA frameworks produce graduates capable of responsible innovation.

Education becomes the engine of civilization evolution.

Pillar 3 - Policy Influence
Rather than lobbying, ZHA contributes intellectual guidance.

Activities include:

- white papers,
- advisory dialogues,
- governance recommendations.

Example: Governments designing national AI strategies incorporate sustainability principles informed by ZHA thought leadership.

Policy evolves proactively.

Pillar 4 - Industry Transformation Leadership

Organizations require practical guidance for Sustainable AI adoption.

ZHA supports:

- ESG-aligned product development,
- responsible automation,
- human-AI collaboration models.

Example: A corporation redesigns AI deployment strategy to prioritize workforce augmentation instead of displacement.

Thought leadership becomes operational change.

Pillar 5 - Global Collaboration Networks
Ideas spread through collaboration.

ZHA builds networks connecting:

- researchers,
- founders,
- educators,
- policymakers,
- communities.

Thought leadership becomes collective intelligence rather than individual authority.

21.3 Research as Thought Leadership

Research under ZHA serves civilization objectives.

Focus areas include:

- sustainable AI architectures,
- human wellbeing analytics,
- climate intelligence,
- ethical algorithm governance.

Research shifts from academic achievement to societal impact.

Example: Sustainable AI Research Program

Researchers collaborate globally to design energy-efficient AI models.

Outcome:

- reduced carbon footprint,
- shared innovation benefits,
- accelerated climate action.

Research becomes planetary responsibility.

21.4 The Role of Communication in Thought Leadership

Ideas influence civilization only when communicated effectively.

ZHA employs multi-layered communication:

- books and publications,
- global summits,
- think tank dialogues,
- digital platforms,
- community engagement.

Communication transforms philosophy into movement.

Example: A sustainability mindset concept introduced through education spreads into corporate governance and public policy discussions.

Thought leadership propagates organically.

21.5 Building Global Trust Through Thought Leadership

Trust becomes the currency of the AI age.

Public trust declines when technology advances without transparency.

ZHA thought leadership builds trust by:

- prioritizing human dignity,
- promoting inclusivity,
- emphasizing long-term responsibility.

Organizations guided by trust achieve sustainable success.

21.6 Thought Leadership as Distributed Intelligence

The ZHA strategy rejects personality-driven influence.

Instead, it promotes distributed thought leadership.

Participants include:

- students,
- educators,
- researchers,
- professionals,
- institutions.

Everyone becomes a contributor to civilization wisdom.

Example: ZSPC Ecosystem

Students conducting sustainability projects contribute ideas influencing organizational innovation practices.

Thought leadership emerges from grassroots participation.

21.7 The Global Thought Leadership Ecosystem

ZHA integrates multiple platforms:

- ZHA Global Think Tank,
- ZHA World Council,
- ZHA Global Institute,
- ZHA World Summit,
- ZSPC Education Ecosystem.

Together they form a continuous intellectual ecosystem guiding global transformation.

21.8 Thought Leadership and Responsible AI Governance

Responsible AI governance requires intellectual alignment before regulation.

ZHA thought leadership helps societies answer:

- What is ethical intelligence?
- What defines sustainable innovation?
- How should AI serve humanity?

Governance becomes guided by shared understanding rather than crisis response.

21.9 From Innovation Leadership to Civilization Leadership

The AI era demands leaders capable of thinking beyond organizational success.

Civilization leadership considers:

- future generations,
- planetary sustainability,
- social harmony,
- ethical intelligence evolution.

Thought leadership becomes civilization leadership.

Example: Sustainable Smart Cities

City planners guided by ZHA philosophy design AI-enabled cities prioritizing:

- environmental resilience,

- community wellbeing,
- inclusive mobility.

Technology serves humanity holistically.

21.10 The ZHA Insight - Ideas Shape the Future of Intelligence
Technological revolutions are remembered not only for inventions but for the philosophies guiding them.

The AI era requires a guiding philosophy:

- Sustainable AI with Human-Centric Purpose.

The ZHA Global Thought Leadership Strategy ensures that:

- intelligence evolves responsibly,
- innovation strengthens civilization,
- humanity remains at the center of technological progress.

Chapter Reflection

Artificial Intelligence will transform civilization.

The decisive question is not whether humanity will innovate but how humanity will think about innovation.

Through global thought leadership, ZHA aligns research, education, governance, and industry toward a shared vision of sustainable progress.

Thought leadership becomes the compass guiding humanity through the Intelligent Age.

The future belongs not to those who build the most powerful technologies, but to those who guide intelligence with wisdom, sustainability, and purpose.

CHAPTER XXII

Chapter 22: Protecting Civilization through Policy and Collaboration

"Sustainable Governance for a Human-Centric Intelligent Civilization"

Introduction - Civilization Protection in the Age of Intelligence

Human civilization has historically been protected through institutions, laws, and shared cultural values.

Laws protected societies from chaos.

Institutions protected economies from instability.

International cooperation protected global peace.

The Artificial Intelligence era introduces a new responsibility:

Humanity must now protect civilization itself from unintended consequences of intelligence.

AI systems increasingly influence:

- economic structures,
- democratic processes,
- education systems,
- environmental sustainability,
- human identity and behavior.

Unlike previous technologies, AI evolves continuously and autonomously.

Therefore, civilization protection can no longer rely solely on reactive regulation.

It requires anticipatory policy and global collaboration.

This chapter explores how Sustainable AI governance, guided by ZHA thought leadership enables humanity to protect civilization while advancing innovation.

22.1 From Reactive Regulation to Anticipatory Governance

Traditional governance responds after harm occurs.

ZHA introduces anticipatory governance.

Instead of asking:

- How do we regulate after failure?

Leaders ask:

- How do we design systems that prevent harm before deployment?

Anticipatory Governance Principles

Ethical impact assessment before innovation.

ESG evaluation embedded in SDLC.

Multi-stakeholder participation.

Continuous monitoring and learning.

Governance evolves from enforcement to stewardship.

Example: Autonomous Transportation

Reactive model:
Deploy vehicles → address accidents later.

Anticipatory model:

- assess urban impact,
- plan workforce transition,
- ensure accessibility inclusion.

Policy guides innovation responsibly.

22.2 Policy as Civilization Infrastructure

In the Sustainable AI era, policy is no longer administrative.

Policy becomes civilization infrastructure.

Well-designed policies:

- enable innovation,
- protect society,
- maintain trust.

Poor policies either:

- suppress innovation, or
- allow uncontrolled disruption.

ZHA proposes balanced governance integrating innovation and sustainability.

Policy Objectives under ZHA

Encourage responsible innovation.

Protect environmental sustainability.

Promote social inclusion.

Maintain governance transparency.

Policy becomes an enabler of human progress.

22.3 The Role of Global Collaboration

No nation can govern AI alone.

AI systems transcend borders instantly.

Therefore collaboration becomes essential.

Global collaboration aligns:

- research,
- regulation,
- education,
- industry practices.

Example: Climate Intelligence Collaboration

Countries sharing AI climate models:

- improve prediction accuracy,
- reduce disaster impact,
- accelerate global response.

Collaboration protects civilization collectively.

22.4 Multi-Stakeholder Policy Collaboration Model

The ZHA framework promotes collaborative governance involving:

Governments
Provide legitimacy and public accountability.

Industry
Implements responsible innovation.

Academia
Generates knowledge and foresight.

Civil Society
Represents human values and societal impact.

Global Institutions
Coordinate international alignment.

Decision-making becomes shared intelligence rather than centralized authority.

22.5 ESG Policies as Civilization Protection Mechanism

Environmental, Social, and Governance frameworks evolve into civilization safeguards.

Environmental Protection

Policies promote:

- energy-efficient AI,
- sustainable data centers,
- circular technology ecosystems.

Example:
Governments incentivize low-carbon AI infrastructure.

Outcome:
Innovation grows while emissions decline.

Social Protection

Policies ensure:

- workforce reskilling,
- inclusive access to AI education,
- digital equity.

Example: Automation tax incentives linked to employee retraining programs.

Technology strengthens society rather than displacing it.

Governance Protection

Policies require:

- algorithm transparency,
- auditability,
- accountability mechanisms.

Example:
Public sector AI decisions must be explainable.

Trust becomes institutionalized.

22.6 Thought Leadership as Policy Catalyst

Policies rarely emerge spontaneously.

They follow intellectual leadership.

ZHA thought leadership influences policy through:

- research publications,
- global dialogues,

- World Summit collaboration,
- Think Tank recommendations.

Ideas shape governance before regulation appears.

Example: ZHA ESG frameworks adopted by industries gradually inform national AI strategies.

Thought leadership becomes policy evolution.

22.7 Collaborative Innovation Agreements

Future civilization governance relies on voluntary cooperation agreements.

Organizations collaborate on:

- sustainability standards,
- ethical AI principles,
- shared risk management.

Competition shifts toward responsible collective advancement.

Example: Open Sustainability Standards

Technology companies agree on shared environmental benchmarks for AI training models.

Global sustainability improves faster than isolated regulation could achieve.

22.8 Protecting Humanity's Future Workforce

One of the greatest AI-era risks is disruption of human purpose.

Policy collaboration focuses on:

- lifelong learning ecosystems,
- human-AI collaboration models,

- dignity-centered employment transitions.

Example: A nation integrates ZHA sustainability mindset education into schools and professional programs.

Workers transition from routine labor to creative and analytical roles.

Civilization adapts rather than fractures.

22.9 The Role of the ZHA Global Institutions

ZHA institutions act as coordination mechanisms:

- ZHA Global Think Tank → research guidance.
- ZHA World Council → foresight governance.
- ZHA Global Institute → leadership education.
- World Summit → collaborative alignment.

Together they create a policy ecosystem protecting civilization continuity.

22.10 Civilization Protection Through Continuous Learning

AI governance cannot remain static.

Civilization protection requires continuous adaptation.

The governance cycle becomes:

- Innovation emerges.
- Impacts assessed.
- Policies adjusted.

Learning integrated globally.

Civilization becomes a learning system.

22.11 Sustainable AI as Global Peace Infrastructure

Historically, cooperation reduced conflict.

In the AI era, shared intelligence systems can become peace infrastructure.

Collaborative AI initiatives:

- reduce resource conflicts,
- support disaster response,
- improve healthcare access,
- strengthen global cooperation.

Technology becomes a force for stability rather than competition.

22.12 The ZHA Insight - Collaboration Protects Civilization

Humanity's greatest strength is cooperation.

The AI era amplifies both:

- risk,
- opportunity.

Civilization will not be protected by technology alone.

It will be protected by:

- shared values,
- responsible policy,
- collaborative governance,
- sustainable innovation.

The ZHA framework ensures that intelligence advances alongside wisdom.

Chapter Reflection

Artificial Intelligence represents humanity's most powerful creation.

Its impact will define the future of civilization.

The question is not whether humanity will innovate.

The question is whether humanity will collaborate wisely enough to protect what innovation transforms.

Through policy alignment and global collaboration, Sustainable AI with Human-Centric Purpose becomes the foundation of civilization protection.

Innovation guided by collaboration sustains civilization.

And civilization guided by sustainability secures humanity's future.

PART VI — The Future Civilization Vision

CHAPTER XXIII

Chapter 23: ZHA 2035 Civilization Roadmap

Designing Humanity's Next Evolution through Sustainable AI with Human-Centric Purpose

Introduction — Humanity at a Civilizational Turning Point

Human civilization stands at a decisive historical moment.

For centuries, progress was measured through:

- industrial expansion,
- economic growth,
- technological advancement.

Yet the 21st century reveals a paradox:

Human capability has reached extraordinary levels,
but planetary stability, social harmony, and human wellbeing remain fragile.
Artificial Intelligence accelerates this paradox.

AI can:

- cure diseases,
- solve climate challenges,
- democratize knowledge,
- transform economies.

But without direction, it can also:

- deepen inequality,
- destabilize employment,
- concentrate power,

- disconnect humanity from purpose.

The future therefore depends not on technology itself, but on how civilization evolves alongside intelligence.

The ZHA 2035 Civilization Roadmap defines a long-term transformation model where:
Technology, sustainability, education, governance, and human consciousness evolve together.

23.1 Why Civilization Needs a Roadmap

Civilizations rarely collapse because of lack of innovation.

They decline when progress loses direction.

The AI era introduces unprecedented speed:

- innovation cycles shorten,
- societal change accelerates,
- institutions struggle to adapt.

Without a roadmap:

- transformation becomes chaotic,
- societies react instead of guiding change.

The ZHA 2035 Roadmap provides:

- shared vision,
- coordinated evolution,
- sustainable progress pathways.

It shifts humanity from technological acceleration to civilizational alignment.

23.2 The ZHA Civilization Vision for 2035

The roadmap envisions a world where intelligence strengthens humanity rather than overwhelms it.

By 2035, civilization evolves toward:

- Sustainable AI integrated into everyday systems,
- Education centered on sustainability mindset,
- Economies aligned with planetary limits,
- Governance guided by foresight,
- Human wellbeing prioritized alongside innovation.

Civilization becomes consciously designed.

Core Civilizational Principle
Progress = Intelligence × Responsibility × Sustainability Mindset

23.3 The ZHA Transformation Timeline (2025 - 2035)

The roadmap unfolds across three phases.

Phase 1 - Alignment (2025 - 2027)

Focus:

- awareness building,
- sustainability mindset education,
- initial AI governance frameworks.

Key Actions:

- ZSPC global expansion,
- Sustainable AI standards introduction,

- Think Tank and World Council activation.

Outcome:
Shared global understanding.

Phase 2 - Integration (2028–2031)

Focus:

- institutional transformation,
- ESG-aligned SDLC adoption,
- collaborative innovation ecosystems.

Key Actions:

- governments align policies,
- industries adopt Sustainable AI practices,
- universities integrate interdisciplinary curricula.

Outcome:
Responsible intelligence embedded in systems.

Phase 3 - Civilization Stewardship (2032–2035)

Focus:

- coordinated global governance,
- civilization wellbeing measurement,
- sustainable innovation scaling.

Key Actions:

- Civilization Index adoption,
- global collaboration networks,
- human-centric intelligent societies.

Outcome:
Intelligence aligned with civilization stability.

23.4 The Role of ZHA Institutions in the Roadmap

The roadmap is operationalized through interconnected institutions.

Together they form humanity's coordination architecture for intelligent civilization.

23.5 AI Transformation Becomes Human Evolution

Traditional digital transformation focused on productivity.

ZHA reframes transformation as:

Human Evolution Through Responsible Intelligence.

AI enhances:

- creativity,
- learning,
- collaboration,
- decision-making.

Humans evolve alongside machines rather than being replaced by them.

Example: Healthcare Evolution

AI assists doctors with predictive diagnostics.

Outcome:

- improved patient care,
- reduced burnout,
- preventive healthcare models.

Technology augments humanity.

23.6 Measuring Civilization Progress - The 2035 Civilization Index

The roadmap introduces new measures of success:

- environmental sustainability,
- social inclusion,
- governance transparency,
- ethical innovation,
- human wellbeing.

GDP alone no longer defines progress.

Civilization health becomes the primary metric.

23.7 Global Collaboration as the Engine of 2035

The roadmap recognizes a fundamental truth:

No nation can achieve sustainable civilization alone.

Global collaboration becomes permanent infrastructure.

Countries collaborate on:

- climate intelligence,
- ethical AI standards,
- education exchange,
- sustainability research.

Humanity functions as an interconnected civilization.

23.8 The Human-Centric Intelligent Civilization

By 2035, ZHA envisions a civilization where:

- AI decisions remain explainable,
- innovation serves humanity,
- ecosystems are protected,
- societies remain inclusive,
- individuals experience purpose and dignity.

Intelligence becomes a partner in human flourishing.

23.9 The ZHA Insight - Civilization Is a Choice
Technology does not determine humanity's future.

Human choices do.

The ZHA 2035 Civilization Roadmap represents a conscious decision:

- to guide intelligence wisely,
- to align progress with sustainability,
- to evolve civilization intentionally.

The future is not something humanity inherits. It is something humanity designs together.

Chapter Reflection

The AI era is not merely a technological revolution.

It is the beginning of a new civilization phase.

The ZHA 2035 Civilization Roadmap transforms Sustainable AI with Human-Centric Purpose into a practical global pathway.

Where technology evolves responsibly.
Where education nurtures wisdom.
Where governance anticipates change.

Where economies sustain life.
Where humanity advances consciously.

The roadmap invites humanity to move beyond survival toward civilization stewardship.

The question before humanity is no longer: Can we build intelligent systems?

The question is: Can we become a civilization wise enough to guide them?

CHAPTER XXIV

Chapter 24: ZHA Civilization Operating System (Z-COS)

Defines the long-term global transformation model where technology, sustainability, education, and governance evolve together.

Operating Humanity's Future through Sustainable AI with Human-Centric Purpose

Introduction - Civilization Needs an Operating System

Every complex system requires an operating system.

Computers require operating systems to coordinate hardware and software.
Organizations require operating models to coordinate people and processes.
Nations require governance systems to coordinate societies.

The Artificial Intelligence era introduces an unprecedented challenge:

Human civilization itself has become a complex intelligent system.

Technology evolves rapidly.
Societies change continuously.
Governance adapts slowly.
Education struggles to keep pace.

Humanity now operates powerful intelligent tools without a unified coordination mechanism.

The ZHA Civilization Operating System (Z-COS) emerges as a response.

Z-COS is not software.

It is a civilization framework that synchronizes technology, sustainability, education, economy, and governance into one coherent system.

It defines how humanity operates responsibly in the age of intelligent systems.

24.1 Why Civilization Requires an Operating System

Historically, civilization evolved organically.

But AI introduces exponential acceleration.

Without coordination:

- innovation outpaces ethics,
- automation disrupts societies,
- governance becomes reactive,
- sustainability becomes secondary.

Civilization risks fragmentation.

Z-COS introduces structured alignment.

It ensures:

- intelligence evolves responsibly,
- institutions collaborate coherently,
- sustainability becomes operational behavior.

Civilization transitions from reactive survival to intentional operation.

24.2 What Is Z-COS?

The ZHA Civilization Operating System is a meta-framework integrating:

- Sustainable AI,
- Sustainability Mindset,

- ESG Governance,
- Responsible Innovation,
- Civilization Stewardship.

It functions as humanity's coordination layer.

Z-COS Definition

Z-COS = Civilization Governance + Sustainable AI + Human-Centric Values + Collective Intelligence

Z-COS aligns five dimensions simultaneously:

24.3 Architecture of the Z-COS Framework

Z-COS operates through layered architecture similar to an operating system.

Layer 1 - Mindset Kernel (Human Foundation)

The core of Z-COS is the ZHA Sustainability Mindset.

Just as an operating system kernel controls system behavior, mindset controls civilization behavior.

The twelve sustainability principles guide:

- leadership decisions,
- organizational culture,
- innovation ethics.

Example:

An AI product team guided by Z-COS automatically evaluates:

- energy consumption,
- social impact,

- governance transparency, before development begins.

Ethics becomes instinctive rather than enforced.

Layer 2 - Intelligence Layer (Sustainable AI)

AI functions as civilization intelligence infrastructure.

Z-COS ensures AI systems:

- augment humans,
- remain transparent,
- minimize environmental impact,
- support societal wellbeing.

Example: Healthcare AI prioritizes prevention, accessibility, and equity instead of profit optimization alone.

Technology aligns with humanity.

Layer 3 - Institutional Layer (Governance & Collaboration)

Institutions coordinate civilization operations.

Integrated structures include:

- ZHA Global Institute,
- ZHA World Council,
- Global Think Tank,
- Education Ecosystem (ZSPC),
- Industry Consortium.

Institutions function like system services within an operating system.

Layer 4 - Operational Layer (SDLC & Organizational Practices)

Organizations implement Z-COS through:

- Sustainable SDLC methodologies,
- ESG metrics,
- Responsible innovation workflows,
- human-AI collaboration models.

Example:

A company deploying automation must:

- reskill workforce,
- measure ESG outcomes,
- evaluate long-term societal effects.

Operations become civilization-aligned.

Layer 5 — Civilization Interface (Societal Outcomes)

The final layer reflects real-world impact:

- sustainable economies,
- inclusive societies,
- protected ecosystems,
- flourishing communities.

Civilization itself becomes the user interface of Z-COS.

24.4 Z-COS Transformation Model

Z-COS transforms civilization across five stages.

Transformation becomes evolutionary rather than disruptive.

24.5 Z-COS and Sustainable AI Integration

Z-COS operationalizes the principle:

AI WITH HUMAN-CENTRIC PURPOSE

AI decisions must satisfy three conditions:

- Beneficial to humanity
- Sustainable for the planet
- Responsible across generations

Example: Smart Agriculture

AI analyzes soil health, water usage, and climate data.

Outcome:

- improved farmer livelihoods,
- reduced environmental degradation,
- sustainable food systems.

Technology becomes ecological intelligence.

24.6 Z-COS in Education Systems

Education becomes the installation mechanism of civilization software.

Students learn:

- sustainability mindset,
- ethical innovation,
- collaborative leadership.

ZSPC acts as the educational deployment platform for Z-COS.

Example: Children trained in sustainability mindset become future leaders who naturally integrate ESG thinking into industry and governance.

Civilization updates itself generationally.

24.7 Z-COS in Governance Systems

Governments adopt Z-COS principles to shift from reactive regulation to anticipatory governance.

Policies integrate:

- foresight modeling,
- ESG measurement,
- citizen participation,
- global cooperation.

Governance becomes adaptive intelligence.

Example: AI simulations evaluate climate policies before implementation, preventing long-term societal harm.

24.8 Z-COS in Economic Systems

Economic systems transition toward sustainable prosperity.

Organizations measure success through:

- civilization wellbeing,
- environmental regeneration,
- human development.

Profit remains important but becomes aligned with societal benefit.

Example: Companies rewarded not only for revenue growth but also for carbon reduction and workforce wellbeing.

Economy supports civilization continuity.

24.9 Z-COS and Global Collaboration Networks

Z-COS enables civilization-scale collaboration.

Instead of isolated innovation:

- countries share sustainability standards,
- researchers collaborate globally,
- industries adopt cooperative innovation.

Humanity operates as a coordinated intelligent ecosystem.

24.10 Z-COS Governance Cycle

The operating system continuously learns through feedback:

- Innovation created
- ESG impact measured
- Society evaluates outcomes
- Governance adapts
- Systems improve

Civilization becomes self-learning.

24.11 Z-COS as Civilization Resilience Framework

Z-COS protects civilization against major disruptions:

- technological unemployment,
- climate crises,
- digital inequality,
- governance instability.

By aligning mindset, institutions, and intelligence, civilization becomes resilient.

Example: AI Disruption Scenario

Automation threatens employment.

Z-COS response:

- reskilling programs activated,
- education systems adapt,
- industry collaborates with government.

Disruption transforms into opportunity.

24.12 The ZHA Insight - Civilization Must Become Intelligent

Humanity has created intelligent machines.

The next step is creating an intelligent civilization.

Z-COS represents this transition.

It ensures:

- intelligence serves humanity,
- innovation remains sustainable,
- governance becomes wise,
- civilization evolves consciously.

Chapter Reflection

The future will not be defined by the most powerful technology.

It will be defined by the system guiding how technology is used.

The ZHA Civilization Operating System transforms Sustainable AI with Human-Centric Purpose into a living operational reality.

Humanity moves from fragmented progress toward coordinated evolution.

Z-COS is not merely a framework.

It is the operating system for a conscious civilization.

CHAPTER XXV

Chapter 25: Education Transformation and Future Generations

"Preparing Humanity for a Sustainable AI Civilization"

Introduction - Civilization Is Ultimately an Educational Project

Every civilization is sustained not by technology alone, but by what it teaches its children.

History repeatedly demonstrates a fundamental truth:

The Renaissance succeeded because education awakened curiosity.

Industrial societies grew because education produced skilled citizens.

The Digital Age expanded because education created technological literacy.

The Artificial Intelligence era introduces a deeper requirement.

Humanity must now educate individuals not only to use intelligence, but to live responsibly with intelligence.

Technology is evolving faster than generations can adapt.

Without educational transformation:

- innovation becomes disruptive,
- societies become divided,
- governance becomes reactive,
- sustainability becomes secondary.

The ZHA framework recognizes that civilization survival depends on education transformation.

The future of Sustainable AI civilization will be determined in classrooms, not laboratories.

25.1 The ZHA Education Transformation Model

ZHA introduces a holistic education model integrating four dimensions.

1. Knowledge Intelligence

Students learn:

- science,
- technology,
- arts,
- social sciences.

AI becomes a learning companion rather than replacement.

2. Sustainability Mindset Formation

Students internalize the twelve ZHA principles:

- feeling sufficient,
- collaboration,
- long-term thinking,
- environmental stewardship,
- continuous learning.

Mindset becomes the foundation of responsible action.

3. Human Capability Development

Education develops uniquely human strengths:

- empathy,
- creativity,
- critical thinking,
- leadership,
- ethical reasoning.

Humans learn what machines cannot replicate.

4. Civilization Responsibility

Students understand their role within:

- society,
- ecosystems,
- global humanity.

Education creates responsible citizens, not just professionals.

25.2 ZSPC - The Education Engine of Sustainable AI Civilization

The ZHA Sustainability Practitioners Certification (ZSPC) ecosystem operationalizes education transformation.

ZSPC enables:

- experiential learning,
- sustainability projects,
- leadership development,
- community engagement.

Students learn by solving real-world problems.

Example: ZSPC Student Project

Students design a local waste management solution using AI analysis.

Outcomes:

- environmental awareness,
- technological literacy,
- civic responsibility.
- Learning becomes lived experience.

25.3 Lifelong Learning Civilization

The AI era eliminates the concept of one-time education.

ZHA promotes lifelong learning ecosystems.

Education continues across:

- childhood,
- professional life,
- leadership stages,
- retirement contribution.

Learning becomes continuous civilization participation.

Example:
A professional retrains every five years in:

- AI collaboration,
- sustainability practices,
- ethical innovation.
- Society remains adaptable.

25.4 Teachers as Civilization Architects

In the ZHA vision, educators become civilization leaders.

Teachers evolve from information transmitters into:

- mentors,
- ethical guides,
- innovation facilitators,
- sustainability ambassadors.

Teacher empowerment becomes essential civilization investment.

Example: Teachers guide students in community sustainability initiatives rather than purely academic assignments.

Education connects learning with real life.

25.5 Integrating Technology into Education Responsibly

AI transforms education delivery but must remain human-centered.

ZHA proposes balanced integration.

AI supports:

- personalized learning,
- adaptive assessment,
- accessibility inclusion.

Humans retain:

- mentorship,
- moral development,
- social learning.

Example: AI Learning Assistant

An AI tutor adapts lessons to individual learning pace.

Teacher focuses on:

- creativity,
- discussion,
- ethical reflection.

Technology amplifies teaching rather than replacing educators.

25.6 Education and Future Workforce Transformation

Automation changes employment patterns.

Education must prepare individuals for:

- evolving careers,
- interdisciplinary roles,
- human-AI collaboration.

Future careers emphasize:

- creativity,
- sustainability leadership,
- social innovation.

Example:

Instead of training only programmers, education develops:

- AI ethicists,
- sustainability analysts,
- human-AI interaction designers.

Workforce evolves alongside civilization needs.

25.7 Global Education Collaboration

The ZHA framework promotes global education alignment.

Institutions collaborate across borders to:

- share sustainability curricula,
- exchange research,
- develop common ethical standards.

Education becomes globally cooperative rather than nationally isolated.

Example: Students from multiple countries collaborate virtually to solve climate challenges.

Future generations grow with global consciousness.

25.8 Education as Preventive Governance

Many societal crises originate from educational gaps.

Education transformation prevents:

- technological misuse,
- social inequality,
- environmental neglect.

Well-educated citizens reduce governance burden.

Governance shifts from regulation to empowerment.

25.9 The Role of Future Generations

Future generations are not passive recipients of civilization.

They are active designers of it.

ZHA prepares youth to become:

- sustainability practitioners,
- responsible innovators,
- ethical leaders,
- civilization stewards.

Youth participation becomes central to global governance.

Example: Young innovators contribute policy ideas through ZHA platforms influencing national AI strategies.

Generations collaborate rather than inherit problems.

25.10 Intergenerational Responsibility

Sustainable civilization requires cooperation between generations.

Older generations contribute:

- wisdom,
- experience,
- institutional knowledge.

Younger generations contribute:

- creativity,
- adaptability,
- technological fluency.

Civilization evolves through intergenerational partnership.

25.11 Education and the Sustainability Mindset Culture

The ultimate goal of education transformation is cultural evolution.

Humanity transitions toward:

- sufficiency over excess,
- collaboration over competition,
- stewardship over exploitation.

Education becomes the mechanism through which civilization consciousness evolves.

25.12 The ZHA Insight - The Future Is Educated Before It Exists

Every future civilization is first imagined in education.

If education teaches competition without responsibility, civilization fragments.

If education cultivates sustainability mindset, civilization flourishes.

The ZHA Education Transformation ensures that Sustainable AI with Human-Centric Purpose becomes a generational inheritance rather than a temporary initiative.

Chapter Reflection

Artificial Intelligence will reshape humanity's future.

Education will determine whether that future becomes sustainable or unstable.

Through education transformation, humanity prepares future generations not merely to survive technological change, but to guide it wisely.

The classroom becomes the birthplace of civilization evolution.

The students of today become the stewards of tomorrow's intelligent world.

CHAPTER XXVI

Chapter 26: Building a Conscious Civilization

"Where Technology, Sustainability, Education, and Governance Evolve Together"

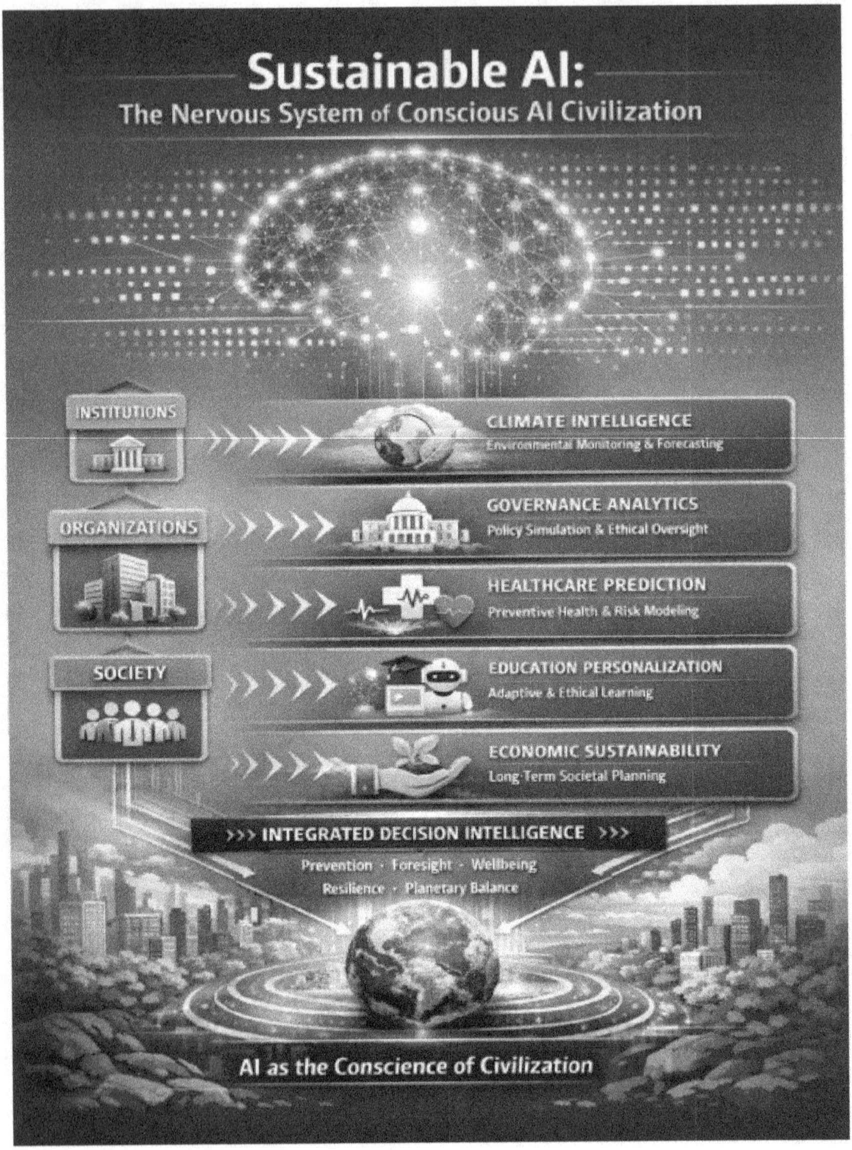

Introduction - Humanity at a Civilizational Threshold

Human history has always been shaped by intelligence.

Fire expanded survival.

Agriculture expanded settlement.

Industry expanded productivity.

Digital technology expanded connectivity.

Artificial Intelligence introduces something fundamentally different.

For the first time, humanity is no longer only using intelligence, humanity is creating intelligence.

This moment represents a civilizational turning point.

The central question of the 21st century is not:

- How powerful will technology become?

but rather:

- How conscious will civilization become while creating intelligence?

Building a conscious civilization means aligning:

- technological power,
- human wisdom,
- planetary sustainability,
- and collective responsibility.

The ZHA framework emerges as a pathway toward this integration.

26.1 What Is a Conscious Civilization?

A conscious civilization is one that understands the consequences of its actions before crises emerge.

It is characterized by:

- awareness before reaction,
- stewardship before exploitation,
- collaboration before domination,
- sustainability before short-term gain.

Conscious civilization does not reject progress.

It guides progress intentionally.

Key Characteristics

A conscious civilization:

- designs technology ethically,
- measures success through wellbeing,
- educates citizens for responsibility,
- governs intelligence collectively,
- protects future generations.

Consciousness becomes civilization's operating principle.

26.2 The Four Pillars of Conscious Civilization

ZHA identifies four interdependent pillars.

1. Technology - Intelligent but Responsible
Technology must evolve from disruption engine to civilization infrastructure.

Sustainable AI ensures:

- transparency,
- fairness,
- ecological responsibility,
- human-centered design.

Technology serves humanity rather than reshaping humanity without consent.

Example: AI healthcare systems prioritize equitable access rather than only profitability.

Innovation becomes humane.

2. Sustainability - Living Within Planetary Limits
Civilization maturity is measured by its relationship with nature.

Sustainability shifts from environmental policy to civilizational value.

A conscious civilization:

- restores ecosystems,
- reduces waste,
- balances consumption,
- respects planetary boundaries.

Economic development aligns with ecological resilience.

Example: Smart cities powered by AI optimize energy, transport, and water usage while improving quality of life.

3. Education - Cultivating Responsible Humans
Education becomes civilization design.

Instead of producing workers alone, education develops:

- thinkers,

- innovators,
- ethical leaders,
- sustainability practitioners.

The ZHA Sustainability Mindset prepares individuals to live responsibly within intelligent systems.

Example: Students learn AI ethics alongside coding, sustainability alongside economics, and leadership alongside knowledge.

4. Governance - Stewardship of Intelligence
Governance evolves from control to stewardship.

Institutions collaborate globally to guide intelligent systems responsibly.

Governance becomes:

- anticipatory,
- participatory,
- transparent,
- globally coordinated.

Civilization governs intelligence rather than reacting to its consequences.

26.3 The Integration Challenge

Historically, these four domains evolved separately:

The AI era demands integration.

Without integration:

- technology accelerates inequality,
- sustainability becomes secondary,
- education becomes outdated,
- governance loses legitimacy.

The ZHA framework integrates all four into one civilization system.

26.4 From Industrial Civilization to Conscious Civilization

Human civilization progresses through stages.

The transition requires humanity to evolve culturally as much as technologically.

Conscious civilization represents humanity's next evolutionary step.

26.5 The Role of Sustainable AI - Institutional & Organizational Transformation

The Core Idea

In the ZHA Conscious Civilization Model, Sustainable AI is not introduced merely as technology adoption.

It becomes:

the nervous system of institutions and organizations

Just as the human nervous system:

- senses reality,
- processes signals,
- predicts danger,
- coordinates response,
- enables learning,

Sustainable AI performs the same role for civilization systems.

Organizations stop reacting to problems after they occur.

They begin anticipating and guiding outcomes.

1. From Automation Tool → Civilization Intelligence Partner

Traditional AI Transformation

Most organizations adopt AI for:

- cost reduction,
- automation,
- productivity improvement,
- competitive advantage.

Result: AI optimizes operations but does not improve civilization outcomes.

ZHA Sustainable AI Transformation

AI becomes embedded into decision intelligence.

Organizations ask:

- What impact will this decision have on society?
- What environmental cost accompanies innovation?
- How does technology influence human wellbeing?

AI evolves into:

- Organizational conscience
- Institutional foresight system
- Civilization guidance mechanism

2. How Sustainable AI Works Inside Institutions
Sustainable AI operates across five institutional intelligence domains.

2.1 Climate Intelligence in Organizations

What Changes
AI continuously monitors environmental impact instead of relying on annual ESG reports.

Institutional Implementation

Universities

- AI tracks campus energy usage.
- Predicts carbon footprint of research labs.
- Optimizes renewable energy consumption.

Corporations

- AI evaluates supply chain emissions in real time.
- Suggests low-carbon production alternatives.

Governments

- AI predicts climate risks before disasters occur.

Example:

A manufacturing organization deploying AI detects:

- excessive energy usage in certain plants.
- predictive models recommend operational redesign.

Outcome:

- emissions reduced,
- costs reduced,
- sustainability achieved simultaneously.

Sustainability becomes operational intelligence.

2.2 Governance Analytics - AI as Institutional Foresight

Traditional governance reacts to crises.

Sustainable AI enables anticipatory governance.

Organizational Practice

AI analyzes:

- policy impact,
- citizen sentiment,
- economic inequality trends,
- ethical risk patterns.

Government Example:

Before implementing urban expansion policy:

AI simulates:

- migration effects,
- employment shifts,
- environmental impact,
- infrastructure stress.

Decision makers adjust policy before harm occurs.

Governance evolves from:
control → stewardship

Corporate Governance Example

AI governance dashboards evaluate:

- algorithm bias risk,
- employee wellbeing,
- ESG compliance,
- ethical incident patterns.

Board decisions become data-guided and human-centered.

2.3 Healthcare Prediction - Preventive Civilization Systems

Healthcare becomes one of the clearest examples of Sustainable AI.

Institutional Transformation

Hospitals move from:
Treatment Model → Prevention Model

AI predicts:

- disease outbreaks,
- chronic illness risk,
- healthcare resource shortages.

Example: National health systems use AI to identify early diabetes risk populations.

Instead of expensive treatment later:

- lifestyle programs introduced early,
- public health cost reduced,
- quality of life improved.

AI becomes civilization health guardian.

2.4 Education Personalization - Building Conscious Citizens

Education institutions become civilization development centers.

Sustainable AI in Education

AI supports:

- personalized learning paths,
- sustainability mindset training,
- ethical reasoning development,
- lifelong learning ecosystems.

ZHA Alignment Example (ZSPC Model)

AI platform tracks:

- student sustainability projects,
- collaboration behavior,
- community impact activities,
- innovation maturity.

Students develop not only skills but responsibility.

Outcome:
Education produces Sustainability Practitioners, not just graduates.

2.5 Economic Sustainability Planning

Economic systems today optimize growth.

Sustainable AI optimizes long-term stability.

Organizational Transformation

AI evaluates:

- workforce displacement risk,
- automation impact,
- social inequality trends,
- future industry shifts.

Example: Corporate AI Adoption

Before automation deployment:

AI simulates:

- jobs affected,
- reskilling requirements,
- regional economic effects.

Company invests in reskilling instead of layoffs.

Result:

- productivity increases,
- social stability preserved,
- organizational reputation strengthened.

Economic intelligence becomes humane intelligence.

3. Sustainable AI Across Organizational Layers
ZHA positions Sustainable AI operating across three layers.

Layer 1 - Operational Intelligence
AI optimizes daily operations.

Examples:

- predictive maintenance,
- resource efficiency,

- workflow automation.

Goal:
Efficiency with sustainability.

Layer 2 - Strategic Intelligence
AI supports leadership decision-making.

Examples:

- ESG forecasting,
- innovation risk analysis,
- long-term investment simulation.

Goal:
Responsible strategy.

Layer 3 - Civilization Intelligence
AI evaluates societal outcomes.

Examples:

- community wellbeing analysis,
- environmental impact forecasting,
- governance transparency measurement.

Goal:
Civilization stability.

4. SDLC Transformation Under Sustainable AI

Sustainable AI reshapes software development itself.

Traditional SDLC Question
Can we build this system?

Sustainable AI SDLC Question
Should we build it, and how does it affect civilization?

Sustainable AI SDLC Stages

Align
define human and ESG outcomes.

Design
ethical architecture decisions.

Develop
bias-aware engineering.

Deploy
monitored societal impact.

Sustain
continuous responsibility lifecycle.

Development becomes stewardship.

5. Institutional Benefits of Sustainable AI

Organizations adopting ZHA Sustainable AI observe:

Operational Benefits

- reduced waste,
- smarter resource use,
- risk prevention.

Cultural Benefits

- ethical awareness,
- collaborative decision-making,

- purpose-driven workforce.

Strategic Benefits

- long-term resilience,
- public trust,
- sustainable innovation leadership.

6. The Shift From Reactive to Preventive Civilization

Historically:
Intelligence becomes preventive civilization infrastructure.

7. Real Integrated Scenario - ZHA Sustainable AI Organization

A ZHA-aligned institution integrates Sustainable AI:

AI monitors environmental footprint.

Governance dashboards track ethical risk.

Employees receive sustainability mindset training.

Education programs cultivate responsible innovators.

Leadership decisions guided by long-term simulations.

Result:

The organization stops behaving like a business unit.

It behaves like a civilization steward.

8. The ZHA Insight

Sustainable AI is not the next technology wave.

It is the operating intelligence of conscious civilization.

When institutions adopt Sustainable AI:

- decisions become wiser,
- innovation becomes humane,
- economies become sustainable,
- governance becomes anticipatory.

AI no longer replaces humans.

AI helps humanity become more responsible than ever before.

26.6 Human Consciousness as the Core Technology

The greatest innovation is not Artificial Intelligence.

It is human awareness.

Technology amplifies intention.

If intention lacks consciousness:

- intelligence accelerates harm.

If intention is conscious:

- intelligence accelerates flourishing.

ZHA's sustainability mindset becomes the psychological foundation of conscious civilization.

26.7 Institutions as Civilization Stabilizers

Conscious civilization requires institutional maturity.

ZHA introduces interconnected institutions:

Global Institute - knowledge and governance alignment.

World Council - foresight and stewardship.

Civilization Charter - shared ethical foundation.

Civilization Index - measuring wellbeing.

Education Ecosystem - preparing future generations.

Institutions transform philosophy into operational civilization.

26.8 Collective Intelligence Over Individual Power

The AI era ends the myth of isolated progress.

No nation, corporation, or individual can govern intelligence alone.

Conscious civilization emerges through collective intelligence.

Collaboration becomes survival strategy.

Example: Global AI safety standards developed collaboratively prevent fragmented technological futures.

Humanity acts as a coordinated species.

26.9 Economic Transformation Toward Wellbeing

Economic systems evolve beyond growth metrics.

Success shifts from GDP to Civilization Wellbeing.

New economic priorities include:

- circular economies,
- regenerative agriculture,
- sustainable innovation,
- human-centered work.

Economies support life rather than consume it.

26.10 Culture as the Ultimate Transformation

Technological revolutions change tools.

Civilizational revolutions change culture.

The deepest transformation of ZHA lies in cultural evolution:

- consumption → sufficiency,
- competition → cooperation,
- ownership → stewardship,
- fear → responsibility.

Culture becomes aligned with sustainability consciousness.

26.11 The Role of Individuals

Conscious civilization is not built only by institutions.

It begins with individuals.

Every person becomes:

- a sustainability practitioner,
- a responsible innovator,
- a participant in governance,
- a steward of future generations.

Civilization transformation scales through human behavior.

26.12 Intergenerational Responsibility

Humanity must shift perspective from short-term success to generational continuity.

Decisions consider:

- children not yet born,
- ecosystems not yet restored,
- technologies not yet created.

Conscious civilization thinks in centuries rather than quarters.

26.13 The ZHA Insight - Conscious Civilization Is a Choice

The future is not predetermined by technology.

It is determined by collective human choice.

Humanity can choose:

intelligence without wisdom, or

intelligence guided by responsibility.

The ZHA framework represents a conscious choice:

- to align intelligence with humanity's highest values.

26.14 The Emerging Future

A conscious civilization will demonstrate:

- Sustainable AI embedded in daily life,
- Education cultivating responsible citizens,
- Governance guided by foresight,
- Economies aligned with planetary balance,
- Technology enhancing human dignity.

Humanity transitions from technological civilization to intelligent civilization.

Final Reflection - The Beginning, Not the End

Civilizations are remembered not for the technologies they created, but for the wisdom with which they used them.

Artificial Intelligence marks humanity's greatest opportunity — and greatest responsibility.

Building a conscious civilization means ensuring that:

- intelligence serves life,
- innovation serves humanity,
- progress serves the planet,
- and civilization serves future generations.

The ZHA vision is therefore not merely a framework.

It is an invitation.

An invitation for humanity to evolve consciously.

An invitation to build a future where intelligence and compassion advance together.

The age of intelligent machines has begun.

The age of conscious civilization must now follow.

Founder Manifesto — Sustainable AI with Human-Centric Purpose

Artificial Intelligence must remain aligned with human dignity, ecological balance, and collective wellbeing. Sustainable AI with Human-Centric Purpose represents a commitment to future generations and the emergence of a conscious civilization.

The Declaration of Sustainable Intelligence

Humanity stands at a defining moment in its evolutionary journey.

For the first time in history, humans are not only creating tools, we are creating intelligence itself.

Artificial Intelligence now participates in decisions once reserved exclusively for human judgment:

- determining medical outcomes,
- influencing democratic discourse,
- shaping economic systems,
- directing environmental management,
- redefining education and human creativity.

This moment demands clarity.

The question before humanity is no longer:

- Can we build intelligent systems?

The question has become:

- What kind of civilization will intelligence serve?

The Foundational Belief

The ZHA Methodology is founded upon a simple but profound conviction:

Artificial Intelligence must remain aligned with:

- Human dignity,
- Ecological balance,
- Collective wellbeing,
- Intergenerational responsibility.

Technology without purpose accelerates disruption.
Technology guided by consciousness enables civilization.

Sustainable AI with Human-Centric Purpose is therefore not merely a technological framework, it is a civilizational commitment.

It represents humanity's promise to future generations that intelligence will advance wisdom, not replace it.

Why Sustainable AI Is Necessary

Earlier technological revolutions transformed:

- physical labor,
- industrial production,
- global communication.

Artificial Intelligence transforms something far deeper:

human cognition itself.

AI influences how people think, decide, learn, trust, and cooperate.

Without intentional alignment:

- efficiency may replace empathy,

- automation may replace meaning,
- intelligence may amplify inequality,
- innovation may outpace responsibility.

The greatest risk of AI is not technological failure.

The greatest risk is ethical misalignment at planetary scale.

Sustainable AI ensures intelligence evolves with humanity, not ahead of it.

Human-Centric Purpose - The Core Principle

Human-centric purpose does not reject progress.

It reframes progress.

AI must:

- augment human potential rather than diminish agency,
- strengthen communities rather than isolate individuals,
- protect ecosystems rather than exploit resources,
- expand opportunity rather than concentrate power.

In this vision:

AI becomes a partner in human flourishing.

From Responsible AI to Sustainable AI

Responsible AI focuses on preventing harm.
Sustainable AI goes further.

Responsible AI protects the present Sustainable AI protects the future.

The Role of AI Transformation Leadership

Technology alone does not transform organizations.

Leadership does.

AI transformation initiatives frequently fail because organizations treat AI as:

- an automation project,
- a cost-reduction initiative,
- a digital modernization effort.

But AI adoption without philosophical alignment creates:

- fragmented innovation,
- ethical risk,
- workforce anxiety,
- ESG instability,
- loss of public trust.

Why AI Transformation Leadership Is Critical

Leaders determine whether AI becomes:

- a productivity tool, or
- a civilization-shaping force.

AI Transformation Leaders act as stewards of intelligence. Their responsibility includes:

Purpose Alignment

- Defining why AI exists within the organization.
- Ensuring AI serves human and societal outcomes that aligns with products development.

Sustainability Mindset Integration

- Embedding ZHA Sustainability Mindset principles into product development.
- Designing energy-efficient AI systems.
- Considering long-term societal consequences.

Human Workforce Evolution

- Reskilling employees instead of replacing them.
- Designing human-AI collaboration models.

Ethical Governance

- Establishing transparent AI decision systems.
- Implementing continuous ethical review processes.

Civilization Responsibility

- Recognizing organizations as contributors to global stability.

AI Transformation Leadership in Institutions

Universities
Leaders ensure students graduate not only as engineers but as responsible innovators.

Example:
AI curricula integrate sustainability mindset, ethics, and societal impact alongside technical skills.

Result:
Future innovators design intelligence responsibly from the beginning.

Corporations
Executives move beyond profit optimization toward sustainable innovation.

Example:
An AI product team chooses optimized models reducing energy consumption instead of maximizing computational scale.

Outcome:
Profitability coexists with planetary responsibility.

Governments
Public leaders use AI for foresight rather than control.

Example:
AI simulations evaluate long-term policy impacts before implementation, preventing economic or environmental harm.

Outcome:
Governance becomes preventive rather than reactive.

Organizations as Civilization Participants

In the AI era, organizations are no longer isolated economic entities.

They become:

- social influencers,
- environmental actors,
- governance participants,
- civilization builders.

AI Transformation Leadership ensures organizations recognize this expanded role.

Every AI deployment becomes a civilizational decision.

The Sustainability Mindset as Leadership Foundation

The ZHA Sustainability Mindset provides the ethical compass for transformation:

Feeling Sufficient prevents technological overconsumption.

Long-Term Thinking protects future generations.

Sharing & Collaboration enables collective intelligence.

Environmental Stewardship safeguards planetary systems.

Continuous Learning allows adaptation in uncertain futures.

Leadership grounded in mindset transforms systems naturally.

Policies become reinforcement, not enforcement.

The Founder's Vision

The ZHA methodology envisions a world where:

- intelligence strengthens humanity,
- innovation respects planetary limits,
- education cultivates conscious leaders,
- governance anticipates consequences,
- technology evolves with wisdom.

This is not a technological ambition.

It is a civilizational aspiration.

A Commitment to Future Generations

Sustainable AI with Human-Centric Purpose represents a promise:

That humanity will not surrender its future to uncontrolled intelligence.

Instead, humanity will:

- guide intelligence with responsibility,
- design innovation with compassion,
- build progress rooted in sustainability.

The emergence of a conscious civilization depends not on smarter machines alone, but on wiser humans leading them.

The ZHA Declaration

We choose:

- Intelligence with responsibility.
- Innovation with sustainability.
- Progress with purpose.

Sustainable AI with Human-Centric Purpose is not merely a framework.

It is the beginning of humanity's next evolutionary chapter.

Conclusion

From Technological Acceleration to Conscious Civilization

Humanity now stands at a civilizational crossroads.

Across history, progress has been defined by humanity's ability to innovate — to invent tools, build machines, and expand technological capability. Each revolution brought unprecedented transformation: agriculture stabilized survival, industry expanded productivity, and digital technologies connected the world.

Yet the Artificial Intelligence era introduces something fundamentally different.

For the first time, humanity is not merely enhancing physical capability.

Humanity is shaping intelligence itself.

This transformation demands a new level of responsibility.

The central challenge of the AI age is not technological advancement — it is civilizational alignment.

The Need for a New Direction

Modern society experiences accelerating technological change:

- algorithms influence human decisions,
- automation reshapes labor,
- digital systems guide governance,
- intelligent machines participate in knowledge creation.

However, civilization has largely approached innovation through acceleration rather than reflection.

Speed has become success.

Scale has become ambition.

Efficiency has become priority.

But acceleration without direction creates instability.

The ZHA Civilization Framework emerges as humanity's response to this imbalance.

It offers a pathway to move beyond technology-driven disruption toward purpose-driven progress.

The ZHA Civilization Framework

The framework presented in this manuscript integrates philosophy, governance, education, innovation, and leadership into a unified model for the Intelligent Age.

At its foundation lies a simple realization:

Technology alone cannot secure humanity's future.
Mindset must guide intelligence.

The ZHA Sustainability Mindset becomes the ethical and cultural operating system enabling Sustainable AI with Human-Centric Purpose.

Through its principles - sufficiency, stewardship, collaboration, long-term thinking, and continuous learning - innovation evolves from competition toward collective wellbeing.

Sustainability ceases to be an external obligation.

It becomes organizational behavior and civilizational identity.

Responsible AI and Sustainable AI - A Unified Path

This manuscript establishes a critical distinction:

- Responsible AI protects society from harm.
- Sustainable AI guides civilization toward long-term flourishing.

www.ingramcontent.com/pod-product-compliance
Lightning Source LLC
LaVergne TN
LVHW052014200926
848481LV00005B/1490